Wind Loads

About the International Code Council®

The International Code Council Inc. (ICC) is a nonprofit association that provides a wide range of building safety solutions including product evaluation, accreditation, certification, codification and training. It develops model codes and standards used worldwide to construct safe, sustainable, affordable and resilient structures. The mission of the Code Council is to provide the highest quality codes, standards, products and services for all concerned with the safety and performance of the built environment. ICC Evaluation Service (ICC-ES) is the industry leader in performing technical evaluations for code compliance fostering safe and sustainable design and construction.

S. K. Ghosh Associates LLC is a member of the ICC Family of Solutions with specialty in seismic and code consulting services providing technical support through publications, seminars, peer reviews, research projects, computer programs, and other support services. www.skghoshassociates.com; (847) 991-2700

ICC Washington DC Headquarters:
500 New Jersey Avenue, NW, 6th Floor, Washington, DC 20001

ICC Regional Offices:
Eastern Regional Office (BIR)
Central Regional Office (CH)
Western Regional Office (LA)
Distribution Center (Lenexa, KS)

888-ICC-SAFE (888-422-7233)
www.iccsafe.org

Wind Loads

Time-Saving Methods Using the 2018 IBC and ASCE/SEI 7-16

David A. Fanella,
Ph.D., S.E., P.E., F.ACI, F.ASCE, F.SEI

New York   Chicago   San Francisco
Athens   London   Madrid
Mexico City   Milan   New Delhi
Singapore   Sydney   Toronto

Library of Congress Cataloging-in-Publication Data

Names: Fanella, David Anthony, author.
Title: Wind loads : time saving methods using the 2018 IBC and ASCE/SEI
   7-16 / David A. Fanella.
Description: [New York] : McGraw Hill Education, [2021] | Includes
   bibliographical references and index. | Summary: "A concise, visual
   guide for engineers designing structures to withstand wind damage, this
   book is part of the new "Time-Saving Methods" series that will present
   explanations and workflows for the 20% of the building code that
   engineers use 80% of the time"—Provided by publisher.
Identifiers: LCCN 2020037101 | ISBN 9781260467420 (paperback) | ISBN
   9781260467437 (ebook)
Subjects: LCSH: Wind-pressure. | Buildings—Aerodynamics. | Wind resistant
   design—Standards—United States.
Classification: LCC TA654.5 .F36 2020 | DDC 624.1/75—dc23
LC record available at https://lccn.loc.gov/2020037101

McGraw Hill books are available at special quantity discounts to use as premiums and sales promotions, or for use in corporate training programs. To contact a representative please visit the Contact Us page at www.mhprofessional.com.

Wind Loads:
Time-Saving Methods Using the 2018 IBC and ASCE/SEI 7-16

1 2 3 4 5 6 7 8 9 CCD 25 24 23 22 21 20

ISBN 978-1-260-46742-0
MHID 1-260-46742-2

Sponsoring Editor
   Ania Levinson

Editorial Supervisor
   Donna M. Martone

Acquisitions Coordinator
   Elizabeth Houde

Project Manager
   Parag Mittal, KnowledgeWorks
   Global Ltd.

Copy Editor
   KnowledgeWorks
   Global Ltd.

Proofreader
   KnowledgeWorks
   Global Ltd.

Production Supervisor
   Pamela A. Pelton

Composition
   KnowledgeWorks
   Global Ltd.

Art Director, Cover
   Jeff Weeks

Contents

Preface

This publication provides structural engineers, educators, students, and other design professionals a concise, visual guide to the determination of structural loads due to wind. The intent is to present the provisions in the 2018 *International Building Code* and ASCE/SEI 7-16 *Minimum Design Loads and Associated Criteria for Buildings and Other Structures* in a manner that is easy to understand and apply. This is achieved by utilizing step-by-step methods including numerous figures, tables, flowcharts, and design aids.

Examples in both inch-pound and metric (S.I.) units illustrate the proper application of the code provisions and follow the step-by-step methods outlined throughout the publication. Section, figure, table, and equation numbers from the code and this publication are given in the right-hand margin of the examples for easy reference.

In short, wind loads can be determined simpler and faster using the procedures in this publication.

For further online information on this topic, please go to https://www.mhprofessional .com/WindLoads

David A. Fanella

About the Author

David. A. Fanella is Senior Director of Engineering at the Concrete Reinforcing Steel Institute where his main responsibility is creating educational material for structural engineers, including publications, design aids, and webinars. He has over 30 years of experience in a wide variety of low-, mid-, and high-rise buildings and other structures and has authored numerous books and technical papers through the years, including two editions of *Reinforced Concrete Structures, Analysis and Design*. David is a licensed Structural Engineer and Professional Engineer in Illinois, and is a Fellow of the American Concrete Institute, the American Society of Civil Engineers, and the Structural Engineers Institute. He is active in many professional organizations, including membership on ASCE/SEI 7 and ACI Committees. He is also past President and past Board Member of the Structural Engineers Association of Illinois.

Wind Loads

CHAPTER 1
Introduction

1.1 Overview

The purpose of this publication is to assist in the proper determination of wind loads in accordance with the 2018 edition of the *International Building Code*© (IBC©) [Ref. 1; see Chap. 7 of this publication for a list of references] and the 2016 edition of ASCE/SEI 7 *Minimum Design Loads and Associated Criteria for Buildings and Other Structures* (Ref. 2). The main goal is to streamline the load determination process by providing straight-forward, step-by-step procedures enhanced by numerous design aids, figures, and flowcharts, which provide a roadmap through the numerous code requirements.

Design professionals will appreciate the simplicity and thoroughness of the content and will find the "how to" methods of load determination useful in everyday practice. Worked-out examples illustrate the proper application of the code requirements and follow the step-by-step procedures noted above; these examples are a valuable resource for individuals studying for licensing exams, undergraduate and graduate students, and others involved in structural engineering.

Readers interested in the background, history, and design philosophy of the code requirements for wind loads can find detailed information and references in the commentary of Ref. 2.

1.2 Scope

Throughout this publication, section numbers from the IBC are referenced as illustrated by the following: Section 1609 of the IBC is denoted as IBC 1609. Similarly, Section 26.1 of ASCE/SEI 7-16 is referenced as ASCE/SEI 26.1.

Chapter 2 contains the parameters required for determining wind loads on the main wind force resisting system (MWFRS) and the component and cladding (C&C) elements of a building or other structure. Information is provided on the basic wind speed, V; wind directionality factor, K_d; exposure; topographic factor, K_{zt}; ground elevation factor, K_e; velocity pressure, q_z; gust-effect factors, G and G_f; enclosure classifications; and, internal pressure coefficients, (GC_{pi}).

The requirements for determining wind pressures and loads on the MWFRS of enclosed, partially enclosed, and open buildings of all heights and various roof configurations in accordance with the Directional Procedure of ASCE/SEI Chapter 27 are covered in Chap. 3.

Chapter 4 contains the requirements for determining wind pressures and loads on the MWFRS of enclosed, partially enclosed, and open low-rise buildings in accordance with the Envelope Procedure of ASCE/SEI Chapter 28.

3

The requirements in ASCE/SEI Chapter 29 are covered in Chap. 5. Included are methods to determine wind pressures and loads on building appurtenances (including rooftop structures and rooftop equipment) and other structures of all heights (including solid freestanding walls, solid freestanding signs, tanks, open signs, single-plane open frames, trussed towers, and rooftop solar panels).

Chapter 6 contains the requirements in ASCE/SEI Chapter 30 for determining wind pressures on C&C elements of enclosed, partially enclosed, and open buildings of all heights and various roof configurations.

The references cited in this publication are given in Chap. 7.

Both inch-pound and S.I. units are used throughout this publication, including in the equations, figures, tables, flowcharts, and examples. In the examples, calculations are performed independently using both sets of units; in other words, the calculations are not performed in one set of units and then converted to the other. Thus, in some cases, the numerical results in inch-pound units do not "exactly" convert to the corresponding numerical results in S.I. units or vice versa.

1.3   Notation

A = effective wind area, ft^2 (m^2)

A_f = area of open buildings and other structures either normal to the wind direction or projected on a plane normal to the wind direction, ft^2 (m^2)

A_g = gross area of the wall in which A_o is identified, ft^2 (m^2)

A_{gi} = sum of the gross surface areas of the building envelope (walls and roof) not including A_g, ft^2 (m^2)

A_n = normalized wind area for rooftop solar panels in ASCE/SEI Figure 29.4-7

A_o = total area of openings in a wall that receives positive external pressure, ft^2 (m^2)

A_{og} = total area of openings in the building envelope, ft^2 (m^2)

A_{oi} – sum of the areas of openings in the building envelope (walls and roof) not including A_o, ft^2 (m^2)

A_s = gross area of the solid freestanding wall or solid sign, ft^2 (m^2)

a = width of pressure coefficient zone, ft (m)

B = horizontal dimension of building measured normal to wind direction, ft (m)

$\bar{b}$ = mean hourly wind speed factor in ASCE/SEI Equation (26.11-16) from ASCE/SEI Table 26.11-1

$\hat{b}$ = 3-s gust speed factor from ASCE/SEI Table 26.11-1

C_f = force coefficient to be used in determination of wind loads for other structures

C_N = net pressure coefficient to be used in determination of wind loads for open buildings

C_p = external pressure coefficient to be used in determination of wind loads for buildings

c = turbulence intensity factor in ASCE/SEI Equation (26.11-7) from Table 26.11-1

D = diameter of a circular structure or member, ft (m)

D' = depth of protruding elements such as ribs and spoilers, ft (m)

d_1 = for rooftop solar arrays, horizontal distance orthogonal to the panel edge to an adjacent panel or the building edge, ignoring any rooftop equipment in ASCE/SEI Figure 29.4-7, ft (m)

d_2 = for rooftop solar arrays, horizontal distance from the edge of one panel to the nearest edge in the next row of panels in ASCE/SEI Figure 29.4-7, ft (m)

F = design wind force for other structures, lb (N)

G = gust-effect factor

G_f = gust-effect factor for MWFRS of flexible buildings and other structures

(GC_p) = product of external pressure coefficient and gust-effect factor to be used in determination of wind loads for buildings

(GC_{pf}) = product of the equivalent external pressure coefficient and gust-effect factor to be used in determination of wind loads for MWFRS of low-rise buildings

(GC_{pi}) = product of internal pressure coefficient and gust-effect factor to be used in determination of wind loads for buildings

(GC_{pn}) = combined net pressure coefficient for a parapet

(GC_r) = product of external pressure coefficient and gust-effect factor to be used in determination of wind loads for rooftop structures

(GC_m) = net pressure coefficient for rooftop solar panels, in ASCE/SEI Equations (29.4-4) and (29.4-5)

$(GC_m)_{nom}$ = nominal net pressure coefficient for rooftop solar panels determined from ASCE/SEI Figure 29.4-7

g_Q = peak factor for background response in ASCE/SEI Equations (26.11-6) and (26.11-10)

g_R = peak factor for resonant response in ASCE/SEI Equation (26.11-10)

g_v = peak factor for wind response in ASCE/SEI Equations (26.11-6) and (26.11-10)

H = height of hill, ridge, or escarpment in ASCE/SEI Figure 26.8-1, ft (m)

h = mean roof height of a building or height of other structure; the eave height is used as the mean roof height where the roof angle θ is less than or equal to 10 degrees, ft (m)

h_e = roof eave height at a particular wall, or the average height if the eave varies along the wall, ft (m)

h_p = height to top of parapet in ASCE/SEI Figures 27.5-2 and 30.6-1, ft (m)

h_{pt} = mean parapet height above the adjacent roof surface for use with ASCE/SEI Equation (29.4-5), ft (m)

h_1 = height of a solar panel above the roof at the lower edge of the panel, ft (m)

h_2 = height of a solar panel above the roof at the upper edge of the panel, ft (m)

$I_{\bar{z}}$ = intensity of turbulence from ASCE/SEI Equation (26.11-7)

K_d = wind directionality factor in ASCE/SEI Table 26.6-1

K_e = ground elevation factor

K_h = velocity pressure exposure coefficient evaluated at height $z = h$

K_z = velocity pressure exposure coefficient evaluated at height z

K_{zt} = topographic factor as defined in ASCE/SEI 26.8

K_1 = multiplier in ASCE/SEI Figure 26.8-1 to obtain K_{zt}

K_2 = multiplier in ASCE/SEI Figure 26.8-1 to obtain K_{zt}

K_3 = multiplier in ASCE/SEI Figure 26.8-1 to obtain K_{zt}

L = horizontal dimension of a building measured parallel to the wind direction, ft (m)

L_b = normalized building length, for use with ASCE/SEI Figure 29.4-7, ft (m)

L_h = distance upwind of crest of hill, ridge, or escarpment in ASCE/SEI Figure 26.8-1 to where the difference in ground elevation is half the height of the hill, ridge, or escarpment, ft (m)

L_p = panel chord length for use with rooftop solar panels in ASCE/SEI Figure 29.4-7, ft (m)

L_r = horizontal dimension of return corner for a solid freestanding wall or solid sign from ASCE/SEI Figure 29.3-1, ft (m)

L_z = integral length scale of turbulence, ft (m)

ℓ = integral length scale factor from ASCE/SEI Table 26.11-1, ft (m)

N_1 = reduced frequency from ASCE/SEI Equation (26.11-14)

n_a = approximate lower bound natural frequency from ASCE/SEI 26.11.2, Hz

n_1 = fundamental natural frequency, Hz

P_L = wind pressure acting on leeward face in ASCE/SEI Figure 27.3-8, lb/ft² (N/m²)

P_W = wind pressure acting on windward face in ASCE/SEI Figure 27.3-8, lb/ft² (N/m²)

p = design pressure to be used in determination of wind loads for buildings, lb/ft² (N/m²)

p_{net} = net design wind pressure from ASCE/SEI Equation (30.4-1), lb/ft² (N/m²)

p_{net30} = net design wind pressure for Exposure B at $h = 30$ ft (9.1 m) and $I = 1.0$ from ASCE/SEI Figure 30.4-1, lb/ft² (N/m²)

p_p = combined net pressure on a parapet from ASCE/SEI Equation (27.3-4), lb/ft² (N/m²)

p_s = net design wind pressure from ASCE/SEI Equation (28.5-1), lb/ft² (N/m²)

p_{s30} = simplified design wind pressure for Exposure B at $h = 30$ ft (9.1 m) and $I = 1.0$ from ASCE/SEI Figure 28.5-1, lb/ft² (N/m²)

Q = background response factor from ASCE/SEI Equation (26.11-8)

q = velocity pressure, lb/ft² (N/m²)

q_h = velocity pressure evaluated at height $z = h$, lb/ft² (N/m²)

q_i = velocity pressure for internal pressure determination, lb/ft² (N/m²)

q_p = velocity pressure at top of parapet, lb/ft² (N/m²)

q_z = velocity pressure evaluated at height z abound ground, lb/ft² (N/m²)

R = resonant response factor from ASCE/SEI Equation (26.11-12)

R_B = value from ASCE/SEI Equations (26.11-15a) and (26.11-15b)

R_h = value from ASCE/SEI Equations (26.11-15a) and (26.11-15b)

R_i = reduction factor from ASCE/SEI Equation (26.13-1)

R_L = value from ASCE/SEI Equations (26.11-15a) and (26.11-15b)

R_n = value from ASCE/SEI Equation (26.11-13)

r = rise-to-span ratio for arched roofs

s = vertical dimension of the solid freestanding wall or solid sign from ASCE/SEI Figure 29.3-1, ft (m)

V = basic wind speed obtained from ASCE/SEI Figures 26.5-1A through 26.5-1D and 26.5-2A through 26.5-2D, which corresponds to a 3-s gust speed at 33 ft (10 m) above the ground in Exposure Category C, mi/h (m/s)

V_i = unpartitioned internal volume, ft³ (m³)

$\bar{V}_{\bar{z}}$ = mean hourly wind speed at height $\bar{z}$, mi/h (m/s)

W = width of building in ASCE/SEI Figures 30.3-3, 30.3-5A, and 30.3-5B and width of span in ASCE/SEI Figures 30.3-4 and 30.3-6, ft (m)

W_L = width of a building on its longest side in ASCE/SEI Figure 29.4-7, ft (m)

W_S = width of a building on its shortest side in ASCE/SEI Figure 29.4-7, ft (m)

x = distance upwind or downwind of crest in ASCE/SEI Figure 26.8-1, ft (m)

z = height above ground level, ft (m)

$\bar{z}$ = equivalent height of structure, ft (m)

z_g = nominal height of the atmospheric boundary layer, ft (m)

z_{min} = exposure constant from ASCE/SEI Table 26.11-1

α = 3-s gust-speed power law exponent from ASCE/SEI Table 26.11-1

$\hat{\alpha}$ = reciprocal of α from ASCE/SEI Table 26.11-1

$\bar{\alpha}$ = mean hourly wind-speed power law exponent in ASCE/SEI Equation (26.11-16) from ASCE/SEI Table 26.11-1

β = damping ratio, percent critical for buildings or other structures

γ_c = panel chord factor for use with rooftop solar panels in ASCE/SEI Equation (29.4-5)

γ_E = array edge factor for use with rooftop solar panels in ASCE/SEI Figure 29.4-7 and ASCE/SEI Equations (29.4-4) and (29.4-5)

γ_p = parapet height factor for use with rooftop solar panels in ASCE/SEI Equation (29.4-5)

ε = ratio of solid area to gross area for solid freestanding wall, solid sign, open sign, face of a trussed tower, or lattice structure

$\bar{\varepsilon}$ = integral length scale power law exponent in ASCE/SEI Equation (26.11-9) from ASCE/SEI Table 26.11-1

η = value used in ASCE/SEI Equations (26.11-15a) and (26.11-15b)

θ = angle of plane of roof from horizontal, degrees

λ = adjustment factor for building height and exposure from ASCE/SEI Figures 28.5-1 and 30.4-1

v = height-to-width ratio for solid sign

ω = angle that a solar panel makes with the roof surface in ASCE/SEI Figure 29.4-7, degrees

CHAPTER 2
General Requirements

2.1 Overview

This chapter contains the parameters required for determining wind loads on the main wind force resisting system (MWFRS) and the component and cladding (C&C) elements of a building or other structure.

According to IBC 202 and ASCE/SEI 26.2, the MWFRS of a building or other structure is an assemblage of structural elements (for example, walls, frames, or a combination thereof) assigned to provide support and stability for the overall structure. The system generally receives wind loading from more than one surface. Elements of the building envelope or of the building appurtenances and rooftop structures and equipment not qualifying as part of the MWFRS are defined as C&C elements.

The following parameters are covered in this chapter:

- Basic wind speed, V (Sec. 2.2)
- Wind directionality factor, K_d (Sec. 2.3)
- Exposure (Sec. 2.4)
- Topographic factor, K_{zt} (Sec. 2.5)
- Ground elevation factor, K_e (Sec. 2.6)
- Velocity pressure, q_z (Sec. 2.7)
- Gust-effect factors, G and G_f (Sec. 2.8)
- Enclosure classifications (Sec. 2.9)
- Internal pressure coefficients, GC_{pi} (Sec. 2.10)

2.2 Basic Wind Speed, V

The basic wind speed, V, is determined at the location of a building or other structure based on risk category (see IBC Table 1604.5 and ASCE/SEI Table 1.5-1 for definitions of risk categories).

Basic wind speeds corresponding to 3-s gust speeds at 33 ft (10 m) above ground for Exposure C and different risk categories are given in IBC Figures 1609.3(1) through 1609.3(8) and ASCE/SEI Figures 26.5-1A through 26.5-1D and ASCE/SEI Figures 26.5-2A through 26.5-2D (the figures in the IBC and ASCE/SEI 7 are identical; also, see Sec. 2.4 of this publication for definitions of exposure categories). A summary of the basic wind speed maps is given in Table 2.1.

Location	Figure No.		Risk Category*	Return Period (years)
	IBC	ASCE/SEI 7		
Conterminous U.S.	1609.3(4)	26.5-1A	I	300
Alaska Puerto Rico	1609.3(1)	26.5-1B	II	700
Guam Virgin Islands	1609.3(2)	26.5-1C	III	1,700
American Samoa	1609.3(3)	26.5-1D	IV	3,000
	1609.3(8)	26.5-2A	I	300
	1609.3(5)	26.5-2B	II	700
Hawaii	1609.3(6)	26.5-2C	III	1,700
	1609.3(7)	26.5-2D	IV	3,000

*See IBC Table 1604.5 and ASCE/SEI Table 1.5-1 for definitions of risk categories.

TABLE 2.1   Summary of Basic Wind Speed Maps in the 2018 IBC and ASCE/SEI 7-16

The crosshatched areas on the maps are designated special wind regions where unusual wind conditions exist. The local authority having jurisdiction over the project should be consulted to obtain the local basic wind speed (ASCE/SEI 26.5.2). Estimation of basic wind speeds from regional climatic data is also permitted (see ASCE/SEI 26.5.3).

Instead of using the maps, basic wind speed can be obtained from Refs. 3 or 4. These online tools assist in determining V by entering either the address or the latitude and longitude coordinates of the site. Serviceability wind speeds corresponding to 10-, 25-, 50-, and 100-year mean recurrence intervals are also provided.

2.3   Wind Directionality Factor, K_d

The wind directionality factor, K_d, accounts for the statistical nature of wind flow and the probability of the maximum effects occurring at any particular time for any given wind direction. Values of K_d based on structure type are given in ASCE/SEI Table 26.6-1 (see Table 2.2).

2.4   Exposure

2.4.1   Overview

Wind must be assumed to come from any horizontal direction when determining wind loads on a building or other structure (ASCE/SEI 26.5.1).

Wind pressures are adjusted based on upwind terrain. An exposure category must be determined upwind for each wind direction considered in design based on

Structure Type		K_d
Buildings	MWFRS	0.85
	C&C	0.85
Arched roofs		0.85
Circular domes		1.0*
Chimneys, tanks, and similar structures	Square	0.90
	Hexagonal	0.95
	Octagonal	1.0*
	Round	1.0*
Solid freestanding walls, roof top equipment, and solid freestanding and attached signs		0.85
Open signs and single-plane open frames		0.85
Trussed towers	Triangular, square, or rectangular	0.85
	All other cases	0.95

*$K_d = 0.95$ is permitted for round or octagonal structures with nonaxisymmetric structural systems.

Table 2.2 Wind Directionality Factor, K_d

the ground surface roughness (ASCE/SEI 26.7). A rational way of satisfying this requirement is to assume eight wind directions with four perpendicular to the main axes of the building or other structure and four at 45-degree angles to the main axes (see ASCE/SEI Figure C26.7-8 and Fig. 2.1). The sectors to be used in determining the exposure for a selected wind direction are numbered 1 to 8 in Fig. 2.1.

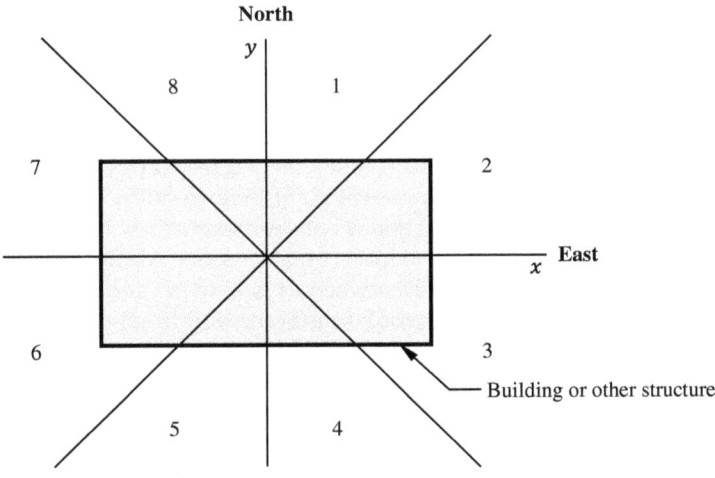

Figure 2.1 Sectors for determining exposure.

Surface Roughness Category	Description
B	Urban and suburban areas, wooded areas, or other terrain with numerous closely spaced obstructions having the size of single-family dwellings or larger
C	Open terrain with scattered obstructions having heights generally less than 30 ft (9.1 m); this category includes flat open country and grasslands
D	Flat, unobstructed areas and water surfaces; this category includes smooth mud flats, salt flats, and unbroken ice

TABLE 2.3 Surface Roughness Categories

Upwind exposure in a particular direction is determined using the surface roughness categories in the 45-degree sectors on each side of the wind direction axis. The sector giving the highest wind loads is used for wind in that direction. For example, for wind blowing north to south, the surface roughness is determined for sectors 1 and 8 and the corresponding exposure resulting in the largest wind loads is used in that direction. For wind along a diagonal, the critical exposure is determined using the sectors on both sides of the diagonal. Full individual wind loading in the x and y directions are determined based on that exposure; 75 percent of these loads are also applied to the building or other structure in each direction at the same time (see ASCE/SEI 27.3.5 and ASCE/SEI Figure 27.3-8 for the design load cases that must be considered).

2.4.2   Surface Roughness Categories

Surface roughness categories are defined in ASCE/SEI 26.7.2 (see Table 2.3). The definitions are descriptive and have been purposely expressed this way for ease in application, while still being sufficiently precise.

2.4.3   Exposure Categories

Exposure categories are determined using the surface roughness categories defined in Table 2.3. Definitions of exposure categories B, C, and D are given in ASCE/SEI 26.7.3 (see Table 2.4).

Upwind surface roughness conditions required for Exposures B and D are illustrated in ASCE/SEI Figures C26.7-1 and C26.7-2, respectively. Aerial photographs for each exposure category are given in ASCE/SEI Figures C26.7-5 through C26.7-7.

For sites located in transition zones between exposure categories, the exposure category resulting in the largest wind loads must be used (ASCE/SEI 26.7.3). For example, the building in Fig. 2.2 is located between Exposures D and B or C. Exposure D must extend the larger of 600 ft (183 m) or 20 times the building height from the point where Exposure D ends (see the definition in Table 2.4). In other words, Exposure D transitions to Exposure B or C over the larger of those two lengths. Thus, Exposure D is required to be used in this transition zone in accordance with ASCE/SEI 26.7.3. However, the exception in that section permits an intermediate exposure category to be used provided it is determined by a rational analysis method defined in recognized literature. An example of such an analysis is given in ASCE/SEI C26.7.

Exposure Category	Definition
B	• Buildings or other structures with a mean roof height $h \le 30$ ft (9.1 m) Surface roughness category B prevails in the upwind direction for a distance > 1,500 ft (457 m) • Buildings or other structures with a mean roof height $h > 30$ ft (9.1 m) Surface roughness category B prevails in the upwind direction for a distance > 2,600 ft (792 m) or 20 times the height of the building or other structure, whichever is greater
C	Applies for all cases where Exposure B or D does not apply
D	• Surface roughness category D prevails in the upwind direction for a distance > 5,000 ft (1,524 m) or 20 times the height of the building, whichever is greater • Surface roughness immediately upwind of the site is B or C, and the site is within a distance of 600 ft (183 m) or 20 times the building or other structure height, whichever is greater, from an Exposure D condition as defined above

TABLE 2.4 Exposure Categories

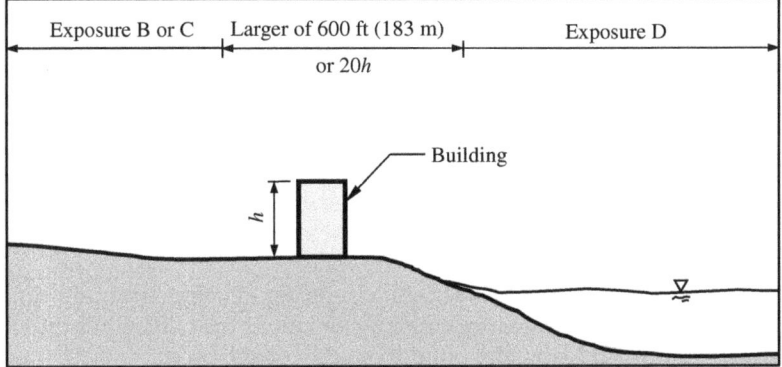

FIGURE 2.2 Transition zone between exposure categories.

A more refined procedure on how to determine the exposure category is given in ASCE/SEI C26.7, including a method on how to account for open patches in a sector.

2.4.4 Exposure Requirements

Exposure requirements pertaining to the wind load procedures in ASCE/SEI 7 are given in ASCE/SEI 26.7.4 (see Table 2.5). Wind loads must be determined based on these requirements.

For the MWFRS of enclosed and partially enclosed buildings and building appurtenances and other structures, wind loads are determined in each direction based on the corresponding exposure in that direction.

Wind Load Procedure	ASCE/SEI 7 Chapter	Requirements
Directional	27	• MWFRS of enclosed and partially enclosed buildings Use an exposure category determined in accordance with ASCE/SEI 26.7.3 in each wind direction • Open buildings with monoslope, pitched or troughed free roofs Use the exposure category determined in accordance with ASCE/SEI 26.7.3 from the eight sectors that results in the highest wind loads for any wind direction
Envelope	28	• MWFRS of all low-rise buildings designed using this procedure Use the exposure category determined in accordance with ASCE/SEI 26.7.3 from the eight sectors that results in the highest wind loads for any wind direction
Directional	29	• Building appurtenances and other structures Use an exposure category determined in accordance with ASCE/SEI 26.7.3 in each wind direction
C&C	30	• C&C Use the exposure category determined in accordance with ASCE/SEI 26.7.3 from the eight sectors that results in the highest wind loads for any wind direction

TABLE 2.5 Exposure Requirements

For C&C elements, open buildings, and low-rise buildings, wind loads are determined based on the exposure from the eight sectors resulting in the largest wind forces for any direction.

The wind loads on the MWFRS of a building or other structure can be determined using only the critical exposure category obtained from all wind directions; this typically results in wind loads that are not overly conservative.

2.5 Topographic Factor, K_{zt}

Buildings and other structures situated on the upper half of isolated hills, ridges, and escarpments can be subjected to larger wind velocities than those located on relatively level terrain. The topographic factor, K_{zt}, accounts for this increase in wind speed and is calculated by ASCE/SEI Equation (26.8-1):

$$K_{zt} = (1 + K_1 K_2 K_3)^2 \qquad (2.1)$$

The topographic multipliers K_1, K_2, and K_3 are determined from ASCE/SEI Figure 26.8-1.

Not every hill, ridge, or escarpment requires an increase in wind velocity; it must be increased only when the site conditions and structure locations of ASCE/SEI 26.8.1 are met. Otherwise, $K_{zt} = 1.0$.

2.6 Ground Elevation Factor, K_e

The ground elevation factor, K_e, adjusts the velocity pressure, q_z, determined in accordance with ASCE/SEI 26.10 based on the reduced mass density of air at elevations above sea level. Values of K_e are given in ASCE/SEI Table 26.9-1; alternatively, K_e can also be calculated using the equations in Note 2 of the table:

$$K_e = e^{-0.0000362 z_g} \qquad (z_g = \text{ground elevation above sea level in ft}) \qquad (2.2)$$

$$K_e = e^{-0.000119 z_g} \qquad (z_g = \text{ground elevation above sea level in m}) \qquad (2.3)$$

It is permitted to take $K_e = 1.0$ for all elevations.

2.7 Velocity Pressure, q_z

The velocity pressure, q_z, at height z above the ground surface is determined by ASCE/SEI Equations (26.10-1) and (26.10-1.si):

$$q_z = 0.00256 K_z K_{zt} K_d K_e V^2 \qquad (\text{lb} / \text{ft}^2) \qquad (2.4)$$

$$q_z = 0.613 K_z K_{zt} K_d K_e V^2 \qquad (\text{N}/\text{m}^2) \qquad (2.5)$$

In these equations, K_z is the velocity pressure exposure coefficient given in ASCE/SEI Table 26.10-1; this coefficient modifies wind velocity (or wind pressure) with respect to height above ground and exposure. Values of K_z can be calculated using the equations in Note 1 at the bottom of ASCE/SEI Table 26.10-1:

$$K_z = \begin{cases} 2.01\left(\dfrac{15}{z_g}\right)^{2/\alpha} & \text{for } z < 15 \text{ ft} \\[3ex] 2.01\left(\dfrac{z}{z_g}\right)^{2/\alpha} & \text{for } 15 \text{ ft} \leq z \leq z_g \end{cases} \qquad (2.6)$$

$$K_z = \begin{cases} 2.01\left(\dfrac{4.6}{z_g}\right)^{2/\alpha} & \text{for } z < 4.6 \text{ m} \\[3ex] 2.01\left(\dfrac{z}{z_g}\right)^{2/\alpha} & \text{for } 4.6 \text{ m} \leq z \leq z_g \end{cases} \qquad (2.7)$$

Values of the 3-s gust speed power law exponent, α, and the nominal height of the atmospheric boundary layer (or, gradient height), z_g, are given in ASCE/SEI Table 26.11-1 based on exposure.

The determination of K_z outlined above is valid for the case of a single roughness category in the direction of analysis (that is, uniform terrain). Procedures on how to determine K_z for a single roughness change or multiple roughness changes are given in ASCE/SEI 26.10.1.

At the mean roof height of a building or other structure, the velocity pressure is q_h, and is determined by Eqs. (2.4) and (2.5) using K_h, which is determined by Eqs. (2.6) and (2.7) at the mean roof height, h.

The flowchart in Fig. 2.3 can be used to determine q_z.

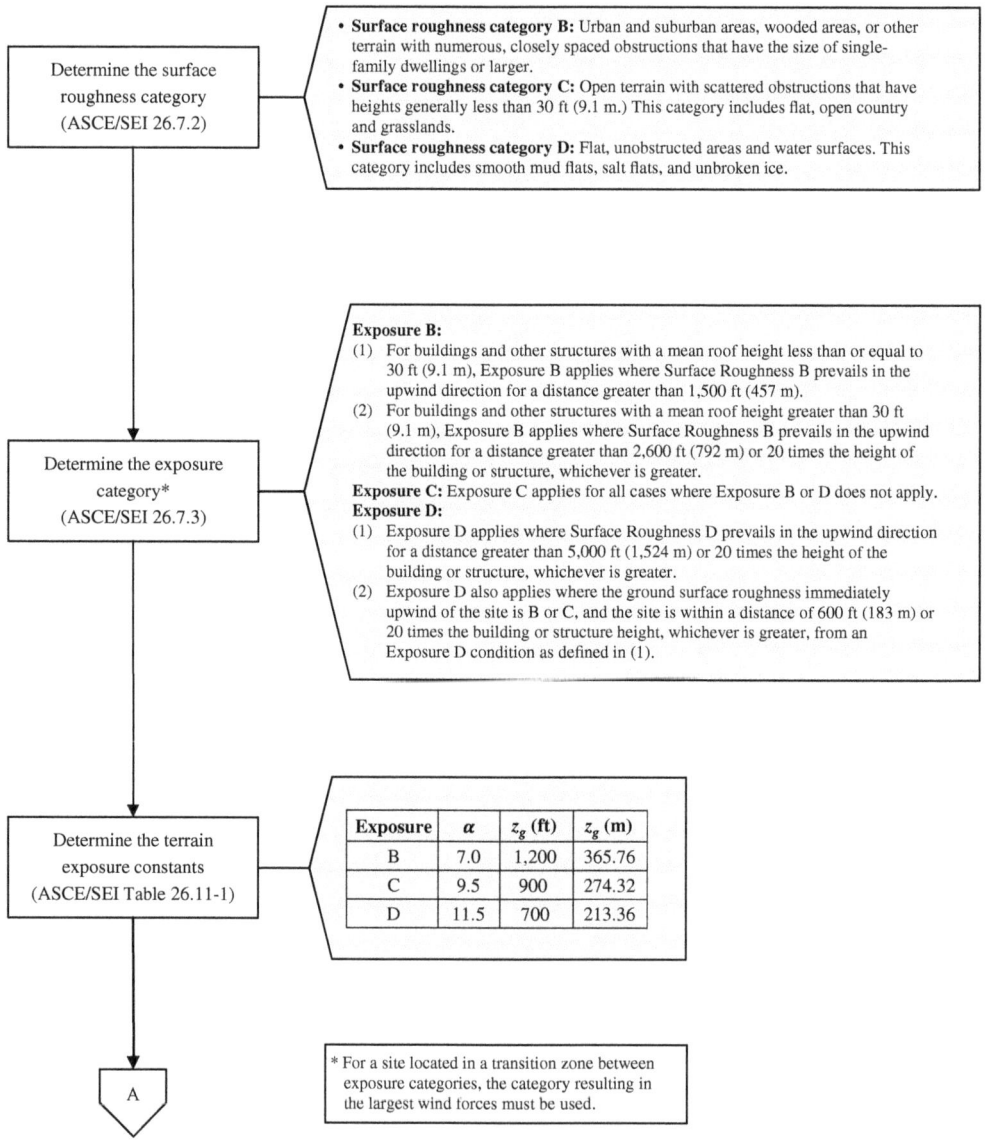

FIGURE 2.3   Flowchart to determine the wind velocity pressure, q_z.

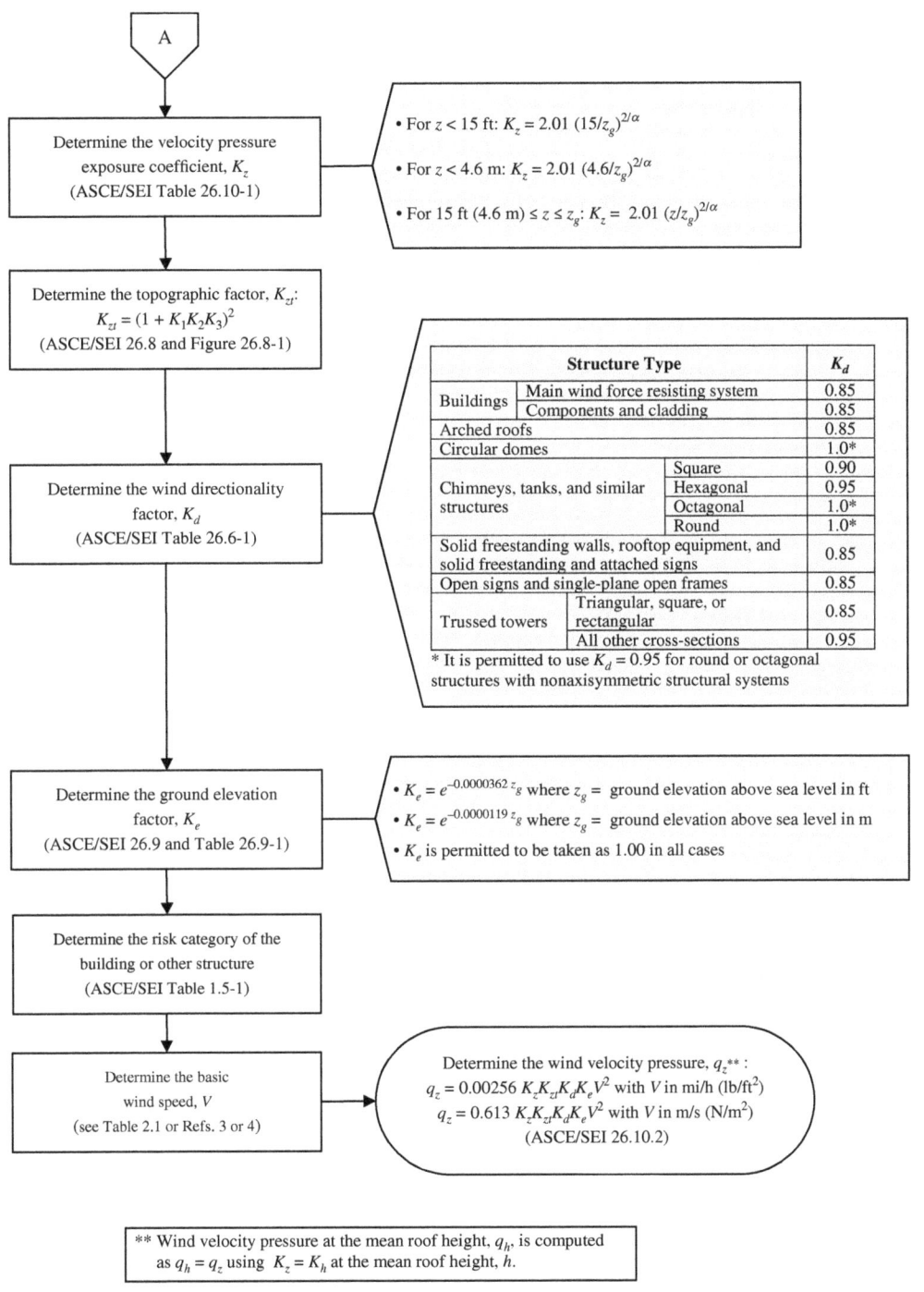

A

Determine the velocity pressure exposure coefficient, K_z (ASCE/SEI Table 26.10-1)

- For $z < 15$ ft: $K_z = 2.01\,(15/z_g)^{2/\alpha}$
- For $z < 4.6$ m: $K_z = 2.01\,(4.6/z_g)^{2/\alpha}$
- For 15 ft $(4.6$ m$) \leq z \leq z_g$: $K_z = 2.01\,(z/z_g)^{2/\alpha}$

Determine the topographic factor, K_{zt}:
$K_{zt} = (1 + K_1 K_2 K_3)^2$
(ASCE/SEI 26.8 and Figure 26.8-1)

Determine the wind directionality factor, K_d
(ASCE/SEI Table 26.6-1)

Structure Type		K_d
Buildings	Main wind force resisting system	0.85
	Components and cladding	0.85
Arched roofs		0.85
Circular domes		1.0*
Chimneys, tanks, and similar structures	Square	0.90
	Hexagonal	0.95
	Octagonal	1.0*
	Round	1.0*
Solid freestanding walls, rooftop equipment, and solid freestanding and attached signs		0.85
Open signs and single-plane open frames		0.85
Trussed towers	Triangular, square, or rectangular	0.85
	All other cross-sections	0.95

* It is permitted to use $K_d = 0.95$ for round or octagonal structures with nonaxisymmetric structural systems

Determine the ground elevation factor, K_e
(ASCE/SEI 26.9 and Table 26.9-1)

- $K_e = e^{-0.0000362\,z_g}$ where $z_g =$ ground elevation above sea level in ft
- $K_e = e^{-0.0000119\,z_g}$ where $z_g =$ ground elevation above sea level in m
- K_e is permitted to be taken as 1.00 in all cases

Determine the risk category of the building or other structure
(ASCE/SEI Table 1.5-1)

Determine the basic wind speed, V
(see Table 2.1 or Refs. 3 or 4)

Determine the wind velocity pressure, q_z** :
$q_z = 0.00256\,K_z K_{zt} K_d K_e V^2$ with V in mi/h (lb/ft^2)
$q_z = 0.613\,K_z K_{zt} K_d K_e V^2$ with V in m/s (N/m^2)
(ASCE/SEI 26.10.2)

** Wind velocity pressure at the mean roof height, q_h, is computed as $q_h = q_z$ using $K_z = K_h$ at the mean roof height, h.

Figure 2.3 (Continued)

2.8   Gust-Effect Factors, G and G$_f$

2.8.1   Overview

The gust-effect factors in ASCE/SEI 26.11 account for the effects of wind gusts on a building or other structure in the along-wind direction.

The natural frequency, n_1, of a structure plays an important role in how the gust-effect factor is determined:

- For rigid buildings and other structures ($n_1 \geq 1$ Hz), the gust-effect factor, G, is determined in accordance with ASCE/SEI 26.11.4.

- For flexible buildings and other structures ($n_1 < 1$ Hz), the gust-effect factor, G$_f$, is determined in accordance with ASCE/SEI 26.11.5.

Approximate methods to determine n_1 are given in ASCE/SEI 26.11.3 (see Sec. 2.8.2 of this publication).

For rigid structures, G is permitted to be taken as 0.85 (ASCE/SEI 26.11.4) or may be determined using ASCE/SEI Equation (26.11-6). Calculating G by this relatively complex equation is usually unnecessary, so G is typically assumed to be 0.85. Low-rise buildings with a mean roof height less than or equal to 60 ft (18.3 m) and less than or equal to the least horizontal dimension of the building are permitted to be considered rigid (ASCE/SEI 26.11.2).

The gust-effect factor for flexible structures, G$_f$, accounts for along-wind loading effects due to dynamic amplification (see ASCE/SEI C26.11) and is determined by ASCE/SEI Equation (26.11-10).

The flowchart in Fig. 2.4 can be used to determine G and G$_f$.

In lieu of the procedures given above, any rational analysis defined in recognized literature is permitted to determine the gust-effect factor (ASCE/SEI 26.11.6).

2.8.2   Natural Frequency, n$_1$

During preliminary design, a sufficient amount of information about the structure may not be available to determine n_1 using computer software. An approximate natural frequency, n_a, can be calculated for buildings using the provisions in ASCE/SEI 26.11.3 provided the height and slenderness conditions in ASCE/SEI 26.11.2.1 are satisfied:

1. Building height must be less than or equal to 300 ft (91.4 m).

2. The building height must be less than 4 times its effective length, L_{eff}, which is determined by ASCE/SEI Equation 26.11-1:

$$L_{eff} = \frac{\sum\limits_{i=1}^{n} h_i L_i}{\sum\limits_{i=1}^{n} h_i} \qquad (2.8)$$

where h_i is the height above grade of level i, L_i is the building length at level i parallel to the wind direction, and n is the number of levels. Where the plan dimension in the direction of analysis does change over the height, L_{eff} is equal to the plan dimension of the building in that direction.

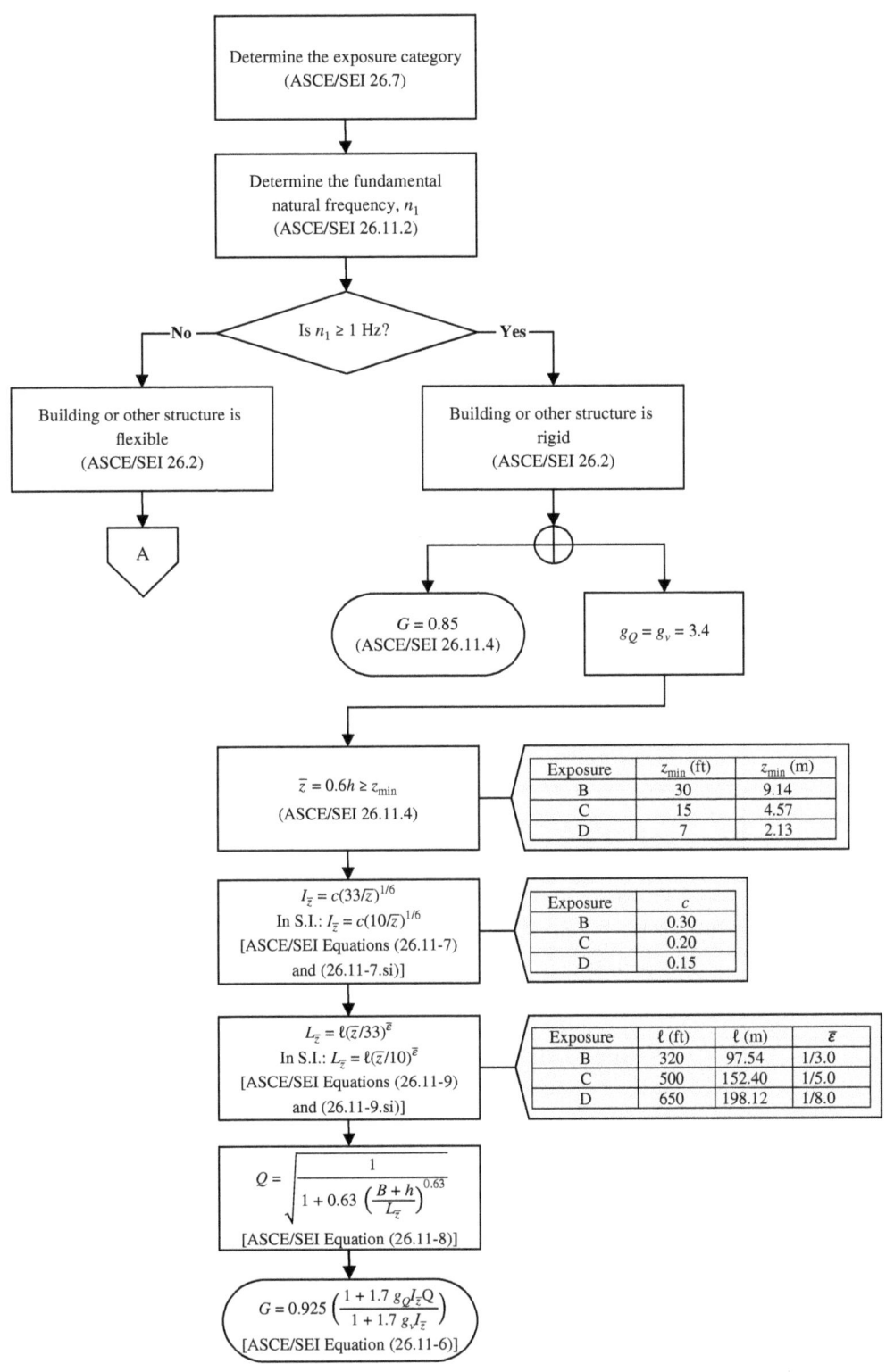

FIGURE 2.4 Flowchart to determine G and G_f.

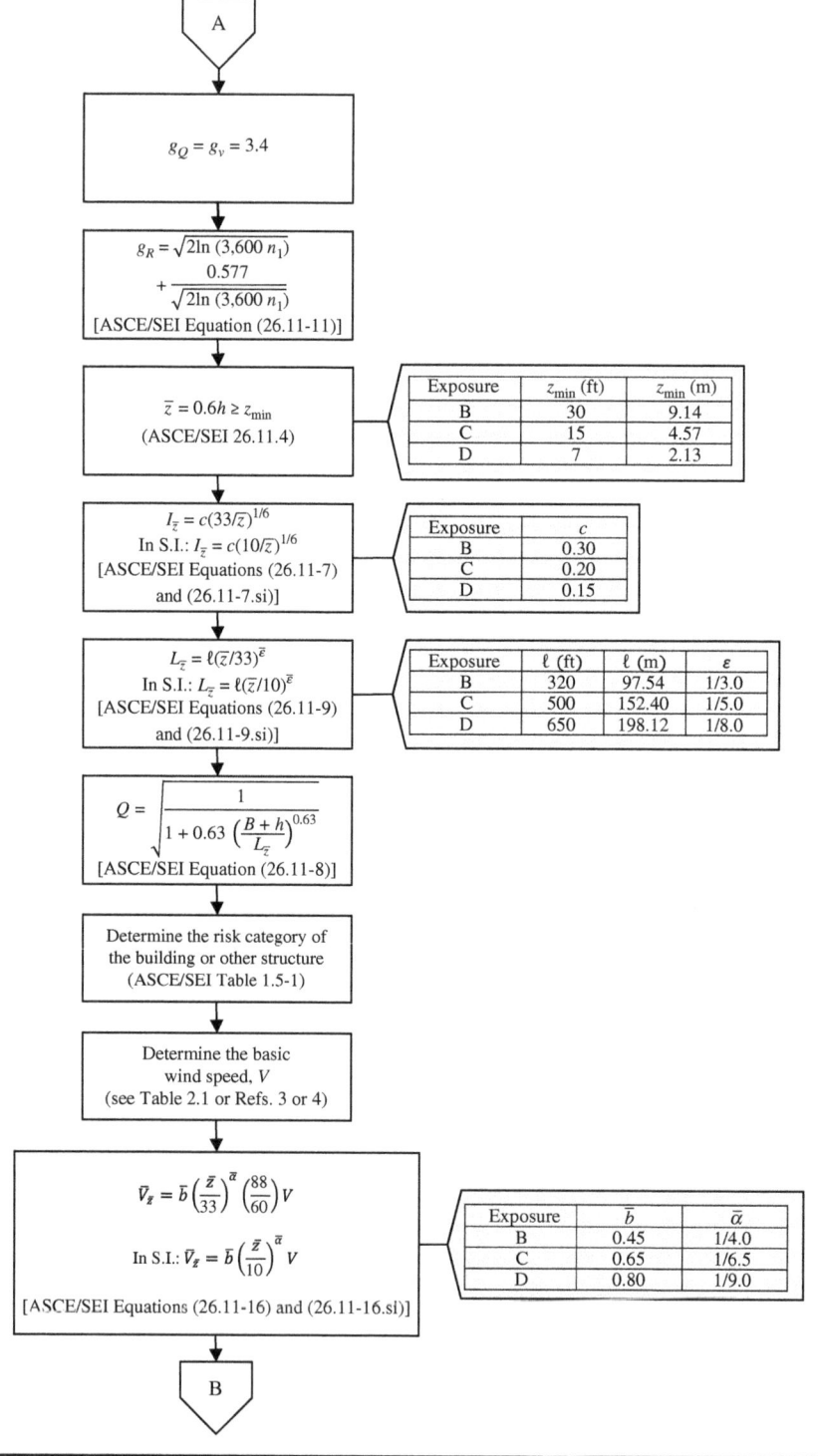

FIGURE 2.4 (Continued)

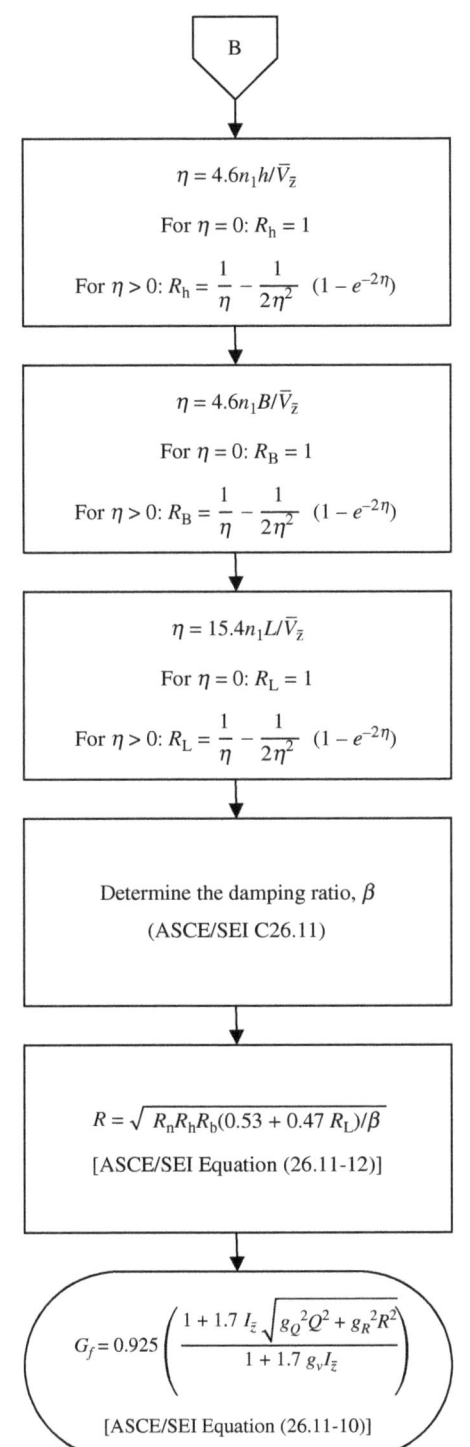

B

$$\eta = 4.6 n_1 h / \bar{V}_{\bar{z}}$$

For $\eta = 0$: $R_h = 1$

For $\eta > 0$: $R_h = \dfrac{1}{\eta} - \dfrac{1}{2\eta^2}\ (1 - e^{-2\eta})$

$$\eta = 4.6 n_1 B / \bar{V}_{\bar{z}}$$

For $\eta = 0$: $R_B = 1$

For $\eta > 0$: $R_B = \dfrac{1}{\eta} - \dfrac{1}{2\eta^2}\ (1 - e^{-2\eta})$

$$\eta = 15.4 n_1 L / \bar{V}_{\bar{z}}$$

For $\eta = 0$: $R_L = 1$

For $\eta > 0$: $R_L = \dfrac{1}{\eta} - \dfrac{1}{2\eta^2}\ (1 - e^{-2\eta})$

Determine the damping ratio, β
(ASCE/SEI C26.11)

$$R = \sqrt{R_n R_h R_b (0.53 + 0.47\ R_L)/\beta}$$

[ASCE/SEI Equation (26.11-12)]

$$G_f = 0.925 \left(\frac{1 + 1.7\ I_{\bar{z}} \sqrt{g_Q^2 Q^2 + g_R^2 R^2}}{1 + 1.7\ g_v I_{\bar{z}}} \right)$$

[ASCE/SEI Equation (26.11-10)]

FIGURE 2.4   *(Continued)*

MWFRS	ASCE/SEI Equation No.	Equations for n_a	
		h in ft	h in m
Structural steel moment-resisting frame buildings	26.11-2	$n_a = \dfrac{22.2}{h^{0.8}}$	$n_a = \dfrac{8.58}{h^{0.8}}$
Concrete moment-resisting frame buildings	26.11-3	$n_a = \dfrac{43.5}{h^{0.9}}$	$n_a = \dfrac{14.93}{h^{0.9}}$
Steel and concrete buildings with lateral force-resisting systems other than moment-resisting frames	26.11-4	$n_a = \dfrac{75}{h}$	$n_a = \dfrac{22.86}{h}$
Concrete or masonry shear wall buildings	26.11-5	$n_a = \dfrac{385(C_w)^{0.5}}{h}$	$n_a = \dfrac{117.3(C_w)^{0.5}}{h}$
		where $$C_w = \frac{100}{A_B} \sum_{i=1}^{n} \left(\frac{h}{h_i}\right)^2 \frac{A_i}{\left[1 + 0.83\left(\dfrac{h_i}{D_i}\right)^2\right]}$$	

TABLE 2.6 Approximate Natural Frequency, n_a

For buildings satisfying the conditions in ASCE/SEI 26.11.2.1, the equations in ASCE/SEI 26.11.3 may be used to determine the approximate natural frequency, n_a (see Table 2.6).

In the equations in Table 2.6, h is the mean roof height of the building. The terms in ASCE/SEI Equation (26.11-5) for concrete or masonry shear walls are defined in ASCE/SEI 26.11.3: n is the number of shear walls in the building in the direction of analysis, A_B is the base area of the building, A_i is the horizontal cross-sectional area of shear wall i, D_i is the length of shear wall i, and h_i is the height of shear wall i.

Additional equations for natural frequency appropriate for wind design are given in ASCE/SEI C26.11.

2.9 Enclosure Classifications

Buildings must be classified as enclosed, partially enclosed, partially open, or open in accordance with the definitions in ASCE/SEI 26.2, which are given in Table 2.7 (ASCE/SEI 26.12.1). The enclosure classification is needed in determining internal pressure coefficients (see Sec. 2.10 of this publication).

The parameters in Table 2.7 are defined as follows (see Fig. 2.5):

- A_o = total area of openings in a wall receiving positive external pressure
- A_g = gross area of wall in which A_o is identified
- A_{oi} = sum of the areas of the openings in the building envelope (walls and roof) not including A_o
- A_{gi} = sum of the gross surface areas of the building envelope (walls and roof) not including A_g

Classification	Definition
Enclosed building	A building complying with the following for each wall: $A_o \leq$ smaller of $\begin{cases} 0.01A_g \\ \\ 4\,ft^2\ \left(0.37\,m^2\right) \end{cases}$
Partially enclosed building	A building complying with both of the following: $\bullet\ A_o > \begin{cases} 1.10A_{oi} \\ \\ \text{smaller of} \begin{cases} 4\,ft^2\ \left(0.37\,m^2\right) \\ \\ 0.01A_g \end{cases} \end{cases}$ $\bullet\ A_{oi}/A_{gi} \leq 0.20$
Partially open building	A building not complying with the requirements for open, partially enclosed, or enclosed buildings
Open building	For each wall in the building, the following is satisfied: $A_o \geq 0.8A_g$

TABLE 2.7 Enclosure Classifications

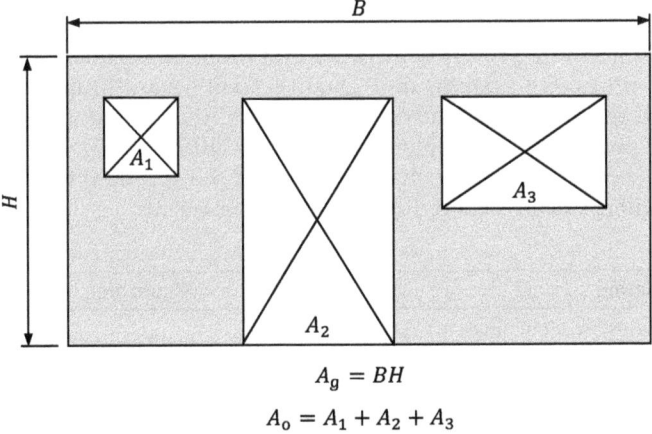

$$A_g = BH$$
$$A_o = A_1 + A_2 + A_3$$

FIGURE 2.5 Definition of wall openings for determination of enclosure classification.

If a building complies with both the open and partially enclosed definitions given in Table 2.7, it must be classified as an open building (ASCE/SEI 26.12.4).

In general, openings are defined as apertures or holes in the building envelope that allow air to flow through the envelope and are designed as open during design winds (ASCE/SEI 26.2). Doors, operable windows, air intake exhaust for air conditioning or ventilating systems, gaps around doors, deliberate gaps in cladding, and flexible and operable louvers are examples of some of the more common types of openings in buildings. "Designed as open" in the definition of openings means the openings are either open during a wind event (like a louver or an air intake exhaust) or can be left open (like doors or operable windows).

In regions that are not prone to hurricanes, it is common practice to assume windows and other glazed units (for example, curtain walls, skylights, and glass doors) are not openings provided these elements have been designed for the applicable C&C wind pressures at the site.

Hurricane-prone regions are located along the U.S. Atlantic Ocean and Gulf of Mexico coasts where $V > 115$ mi/h (51.4 m/s) for Risk Category II buildings. Hawaii, Puerto Rico, Guam, Virgin Islands, and American Samoa are also designated hurricane-prone regions.

Wind-borne debris regions are located in hurricane-prone regions as follows (IBC 202 and ASCE/SEI 26.12.3.1):

- Within 1 mi (1.6 km) of the coastal mean high water line where $V \geq 130$ mi/h (58 m/s)
- In areas where $V \geq 140$ mi/h (63 m/s)

The figures in the IBC and ASCE/SEI 7 to be used in determining wind-borne debris regions based on building classification are given in Table 2.8.

Special requirements are given in IBC 1609.2 and ASCE/SEI 26.12.3.2 for the protection of glazed openings in wind-borne regions. With a few exceptions (see ASCE/SEI 26.12.3.1), all glazing must be protected with an impact-protective system (such as shutters or screens) or the glazing itself must be impact-resistant. The ASTM standards in ASCE/SEI 26.12.3.2 prescribe the tests that must be performed to determine the suitability of protective systems and glazing. Glazing and impact-protective systems in buildings and other structures classified as Risk Category IV must comply with enhanced protection requirements (see ASCE/SEI 26.12.3.2). Impact-protective systems and impact-resistant glazing that have passed the specified tests essentially ensure the glazing will not be breached during a hurricane event.

Classification	Figure Nos.
• Risk Category II buildings and other structures • Risk Category III buildings and other structures, except healthcare facilities	IBC Figures 1609.3(1) and 1609.3(5) ASCE/SEI Figures 26.5-1B and 26.5-2B
Risk Category III healthcare facilities	IBC Figures 1609.3(2) and 1609.3(6) ASCE/SEI Figures 26.5-1C and 26.5-2C
Risk Category IV buildings and other structures	IBC Figures 1609.3(3) and 1609.3(7) ASCE/SEI Figures 26.5-1D and 26.5-2D

TABLE 2.8 Figures in the IBC and ASCE/SEI 7 to be used in Determining Wind-Borne Debris Regions

2.10   Internal Pressure Coefficients

Internal pressure coefficients (GC_{pi}) are given in ASCE/SEI Table 26.13-1 and are based on the enclosure classification determined in accordance with ASCE/SEI 26.12 (see Sec. 2.9 of this publication and Table 2.9). Depicted in Fig. 2.6 are positive and negative internal pressures acting on the interior surfaces of a building.

Enclosure Classification in Accordance with ASCE/SEI 26.12	(GC_{pi})
Enclosed buildings	+0.18
	−0.18
Partially enclosed buildings	+0.55
	−0.55
Partially open buildings	+0.18
	−0.18
Open buildings	0.00

TABLE 2.9   Internal Pressure Coefficients (GC_{pi}) in Accordance with ASCE/SEI 26.13

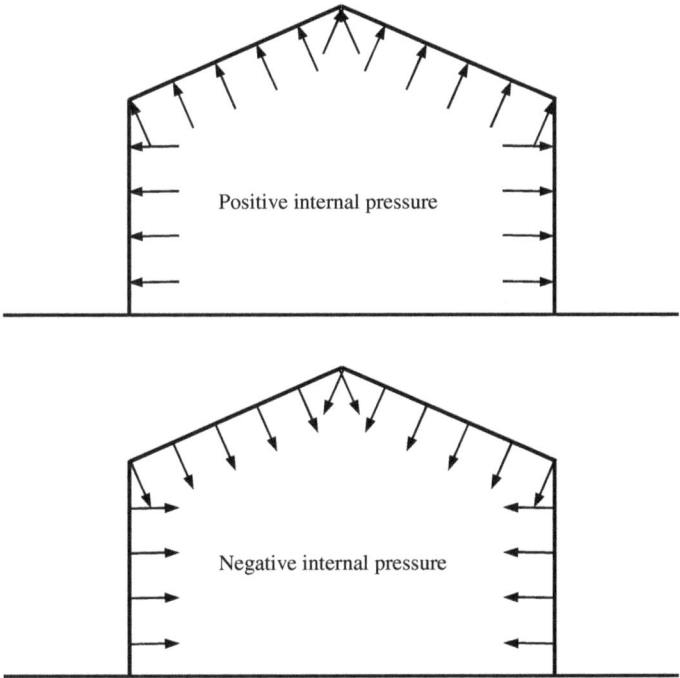

FIGURE 2.6   Positive and negative internal pressures on the interior surfaces of a building.

Gust and aerodynamic effects are combined into one factor (GC_{pi}), and thus the gust-effect factor must not be determined separately in the analysis (ASCE/SEI 26.11.7).

It is evident from Table 2.9 that the value of (GC_{pi}) for partially enclosed buildings is approximately 3 times that for enclosed and partially open buildings. Determining the correct enclosure classification is very important in order to capture the appropriate effect of internal pressure on the design of applicable members in a building.

The internal pressure coefficient may be reduced by the reduction factor R_i given in ASCE/SEI Equation (26.13-1) for partially enclosed buildings containing a single, unpartitioned large volume (ASCE/SEI 26.13.1).

2.11   Examples

The following examples illustrate the determination of topographic factor, K_{zt}; velocity pressure, q_z; gust-effect factors, G and G_f; and enclosure classifications.

2.11.1   Example 2.1—Calculation of Topographic Factor, K_{zt}

Determine the topographic factors, K_{zt}, for the material storage building in Fig. 2.7. The building is located on the downwind side of a two-dimensional escarpment at an Exposure C site in Peoria, IL.

Solution

Step 1—Determine if the conditions in ASCE/SEI 26.8.1 are satisfied

- Assume the topography is such that conditions 1 and 2 are satisfied.
- Condition 3 is satisfied because the building is located near the crest of the escarpment.
- $H/L_h = 20/30 = 0.67 > 0.2$, so condition 4 is satisfied.
- $H = 20$ ft (6.1 m) > 15 ft (4.6 m) for Exposure C, so condition 5 is satisfied.

Because all five conditions are satisfied, wind speed-up effects at the escarpment must be considered in design.

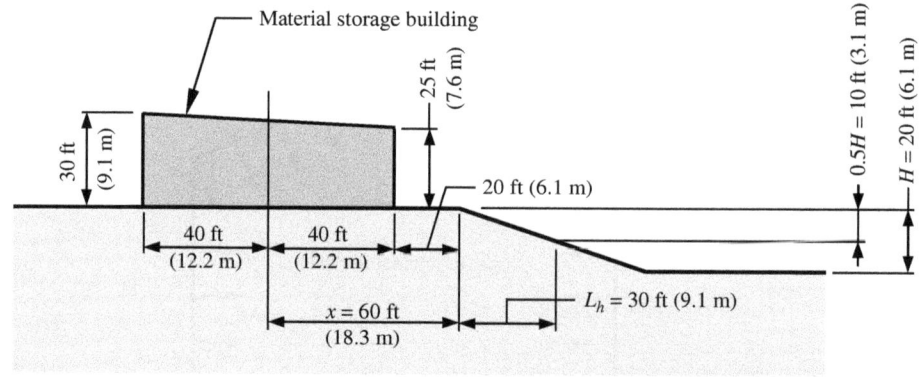

FIGURE 2.7   Material storage building located on an escarpment in Example 2.1.

Step 2—Determine K_{zt} over the height of the building   ASCE/SEI 26.8.2

$$K_{zt} = (1 + K_1 K_2 K_3)^2 \qquad \text{ASCE/SEI Equation (26.8-1)}$$

The multipliers K_1, K_2, and K_3 are determined using ASCE/SEI Figure 26.8-1.
    It was determined in Step 1 that $H/L_h = 0.67$. According to Footnote b in ASCE/SEI Figure 26.8-1, where $H/L_h > 0.50$, use $H/L_h = 0.50$ when evaluating K_1 and substitute $2H$ for L_h when evaluating K_2 and K_3.
    From ASCE/SEI Figure 26.8-1, $K_1 = 0.43$ for $H/L_h = 0.50$, a two-dimensional escarpment, and Exposure C.

$$x/L_h = x/2H = 60/(2 \times 20) = 1.5$$

In S.I.: $x/L_h = x/2H = 18.3/(2 \times 6.1) = 1.5$
    From ASCE/SEI Figure 26.8-1, $K_2 = 0.63$ for a two-dimensional escarpment with $x/L_h = 1.5$.

    When determining K_3, z is taken midway between the height range (it is unconservative to use the top height of the range when determining K_3). For example, assume the upper range is from the eave height [$z = 25$ ft (7.6 m)] to 20 ft (6.1 m). The value of z to use in the determination of K_3 is equal to $(25 + 20)/2 = 22.5$ ft (6.9 m).

    At this height, $z/L_h = z/2H = 22.5/(2 \times 20) = 0.56$.

In S.I.: $z/L_h = z/2H = 6.86/(2 \times 6.10) = 0.56$.
    From ASCE/SEI Figure 26.8-1, $K_3 = e^{-\gamma(z/2H)} = e^{-2.5 \times 0.56} = 0.25$ for a two-dimensional escarpment where the height attenuation factor, γ, is equal to 2.5 for two-dimensional escarpments.

    Thus, the topographic factor, K_{zt}, is equal to the following:

$$K_{zt} = \left[1 + (0.43 \times 0.63 \times 0.25)\right]^2 = 1.14$$

    A summary of the topographic factors over the given height ranges is given in Table 2.10.

2.11.2   Example 2.2—Calculation of Velocity Pressures, q_z

Determine the velocity pressures, q_z, for the material storage building in Fig. 2.7. The building has a mean roof height of 25 ft (7.6 m) and is located on the downwind side of a two-dimensional escarpment at an Exposure C site in Peoria, IL.

Height above Ground Level z	Height z to Be Used in the Determination of K_3	$z/2h$	K_3	K_{zt}
25.0 ft (7.6 m)	22.5 ft (6.9 m)	0.56	0.25	1.14
20.0 ft (6.1 m)	17.5 ft (5.3 m)	0.44	0.34	1.19
15.0 ft (4.6 m)	7.5 ft (2.3 m)	0.19	0.62	1.36

TABLE 2.10   Topographic Factors, K_{zt}, for the Building in Example 2.1

Solution

The flowchart in Fig. 2.3 is used to determine q_z.

> *Step 1—Determine the surface roughness category*   *ASCE/SEI 26.7.2*
>
> The exposure is given as C in the problem statement, so assume surface roughness C is present in all directions.

> *Step 2—Determine the exposure category*   *ASCE/SEI 26.7.3*
>
> The exposure is given as C in the problem statement.

> *Step 3—Determine the terrain exposure constants*   *ASCE/SEI Table 26.11-1*
>
> For Exposure C, $\alpha = 9.5$ and $z_g = 900$ ft (274.32 m).

> *Step 4—Determine the velocity pressure exposure coefficient, K_z*   *ASCE/SEI Table 26.10-1*
>
> Values of K_z and K_h for Exposure C are given in Table 2.11.

> *Step 5—Determine the topographic factors, K_{zt}*   *ASCE/SEI Figure 26.8-1*
>
> The topographic factors are given in Table 2.10 in Example 2.1.

> *Step 6—Determine the wind directionality factor, K_d*   *ASCE/SEI Figure 26.6-1*
>
> The wind directionality factor, K_d, is equal to 0.85 for both the MWFRS and the C&C elements for this building.

> *Step 7—Determine the ground elevation factor, K_e*   *ASCE/SEI Figure 26.9-1*
>
> It is permitted to take $K_e = 1.0$ for all elevations.

> *Step 8—Determine the risk category of the building*   *ASCE/SEI Table 1.5-1*
>
> Assuming this material storage building represents low risk to human life in the event of failure, it is assigned to Risk Category I.

> *Step 9—Determine the basic wind speed, V*
>
> For Risk Category I, use IBC Figure 1609.3(4) or ASCE/SEI Figure 26.5-1A.
>
> Equivalently, use Refs. 3 or 4 to obtain $V = 101$ mi/h (45 m/s) for Peoria, IL.

Height above Ground Level, z	K_z
25.0 ft (7.6 m)	0.94
20.0 ft (6.1 m)	0.90
15.0 ft (4.6 m)	0.85

TABLE 2.11   Velocity Pressure Exposure Coefficients, K_z and K_h

Step 10—Determine the velocity pressures, q_z   *ASCE/SEI 26.10.2*

The velocity pressures, q_z, over the height of the building are given in Table 2.12. These pressures are determined using ASCE/SEI Equations (26.10-1) and (26.10-1.si). For example, the velocity pressure at the eave height of the building is equal to the following:

$$q_h = 0.00256 K_h K_{zt} K_d K_e V^2 = 0.00256 \times 0.94 \times 1.14 \times 0.85 \times 1.0 \times 101^2 = 23.8 \text{ lb/ft}^2$$

In S.I.:

$$q_h = 0.613 K_h K_{zt} K_d K_e V^2 = 0.613 \times 0.94 \times 1.14 \times 0.85 \times 1.0 \times 45^2 / 1,000 = 1.13 \text{ kN/m}^2$$

2.11.3 Example 2.3—Calculation of Gust-Effect Factor, G

Determine the gust-effect factor, G, for the material storage building in Fig. 2.7 assuming the wind direction is normal to the 140-ft (42.7 m) side of the building. The building is located on the downwind side of a two-dimensional escarpment at an Exposure C site in Peoria, IL.

Solution

The flowchart in Fig. 2.4 is used to determine G.

Step 1—Determine the natural frequency, n_1   *ASCE/SEI 26.11.2*

A preliminary analysis indicates $n_1 > 1.0$ Hz.

Step 2—Determine if the building is rigid or flexible   *ASCE/SEI 26.2*

Because $n_1 > 1.0$ Hz, the building is classified as rigid.

Step 3—Determine the gust-effect factor, G, for rigid buildings   *ASCE/SEI 26.11.4*

For rigid buildings, G is permitted to be taken as 0.85.

For comparison purposes, G is determined using ASCE/SEI Equation (26.11-6).

- Step 3a—Determine g_Q and g_v

$$g_Q = g_v = 3.4$$

- Step 3b—Determine $\bar{z}$

$$\bar{z} = 0.6 \; h = 0.6 \times 25.0 = 15.0 \text{ ft} = z_{min} = 15.0 \text{ ft}$$ ASCE/SEI Table 26.11-1

Height above Ground Level, z	K_z	K_{zt}	q_z, lb/ft² (kN/m²)
25.0 ft (7.6 m)	0.94	1.14	23.8 (1.13)
20.0 ft (6.1 m)	0.90	1.19	23.8 (1.13)
15.0 ft (4.6 m)	0.85	1.36	25.7 (1.22)

TABLE 2.12 Velocity Pressures, q_z and q_h

In S.I.:

$$\bar{z} = 0.6\, h = 0.6 \times 7.62 = 4.57 \text{ m} = z_{min} = 4.57 \text{ m}$$

- Step 3c—Determine $I_{\bar{z}}$

$$I_{\bar{z}} = c\left(\frac{33}{\bar{z}}\right)^{1/6} = 0.20 \times \left(\frac{33}{15.0}\right)^{1/6} = 0.2281 \qquad \text{ASCE/SEI Equation (26.11-7)}$$

In S.I.:

$$I_{\bar{z}} = c\left(\frac{10}{\bar{z}}\right)^{1/6} = 0.20 \times \left(\frac{10}{4.57}\right)^{1/6} = 0.2279 \qquad \text{ASCE/SEI Equation (26.11-7.si)}$$

- Step 3d— Determine $L_{\bar{z}}$

$$L_{\bar{z}} = \ell\left(\frac{\bar{z}}{33}\right)^{\bar{\varepsilon}} = 500 \times \left(\frac{15.0}{33}\right)^{1/5} = 427.1 \text{ ft} \qquad \text{ASCE/SEI Equation (26.11-9)}$$

In S.I.:

$$L_{\bar{z}} = \ell\left(\frac{\bar{z}}{10}\right)^{\bar{\varepsilon}} = 152.40 \times \left(\frac{4.57}{10}\right)^{1/5} = 130.31 \text{ m} \qquad \text{ASCE/SEI Equation (26.11-9.si)}$$

- Step 3e—Determine Q

$$Q = \sqrt{\frac{1}{1+0.63\left(\dfrac{B+h}{L_{\bar{z}}}\right)^{0.63}}} = \sqrt{\frac{1}{1+\left[0.63 \times \left(\dfrac{140.0+25.0}{427.1}\right)^{0.63}\right]}} = 0.86$$

$$\text{ASCE/SEI Equation (26.11-8)}$$

In S.I.:

$$Q = \sqrt{\frac{1}{1+0.63\left(\dfrac{B+h}{L_{\bar{z}}}\right)^{0.63}}} = \sqrt{\frac{1}{1+\left[0.63 \times \left(\dfrac{42.67+7.62}{130.31}\right)^{0.63}\right]}} = 0.86$$

- Step 3f—Determine G

$$G = 0.925\left(\frac{1+1.7 g_Q I_{\bar{z}} Q}{1+1.7 g_v I_{\bar{z}}}\right) = 0.925 \times \left[\frac{1+(1.7 \times 3.4 \times 0.2281 \times 0.86)}{1+(1.7 \times 3.4 \times 0.2281)}\right] = 0.85$$

In S.I.:

$$G = 0.925\left(\frac{1+1.7 g_Q I_{\bar{z}} Q}{1+1.7 g_v I_{\bar{z}}}\right) = 0.925 \times \left[\frac{1+(1.7 \times 3.4 \times 0.2279 \times 0.86)}{1+(1.7 \times 3.4 \times 0.2279)}\right] = 0.85$$

2.11.4   Example 2.4—Determination of Enclosure Classification

Determine the enclosure classification for the material storage building in Fig. 2.8. The building has a monoslope roof in the east-west direction with a slope of 3.58 degrees. The north, south, and east walls are open. The west wall has two 20 ft

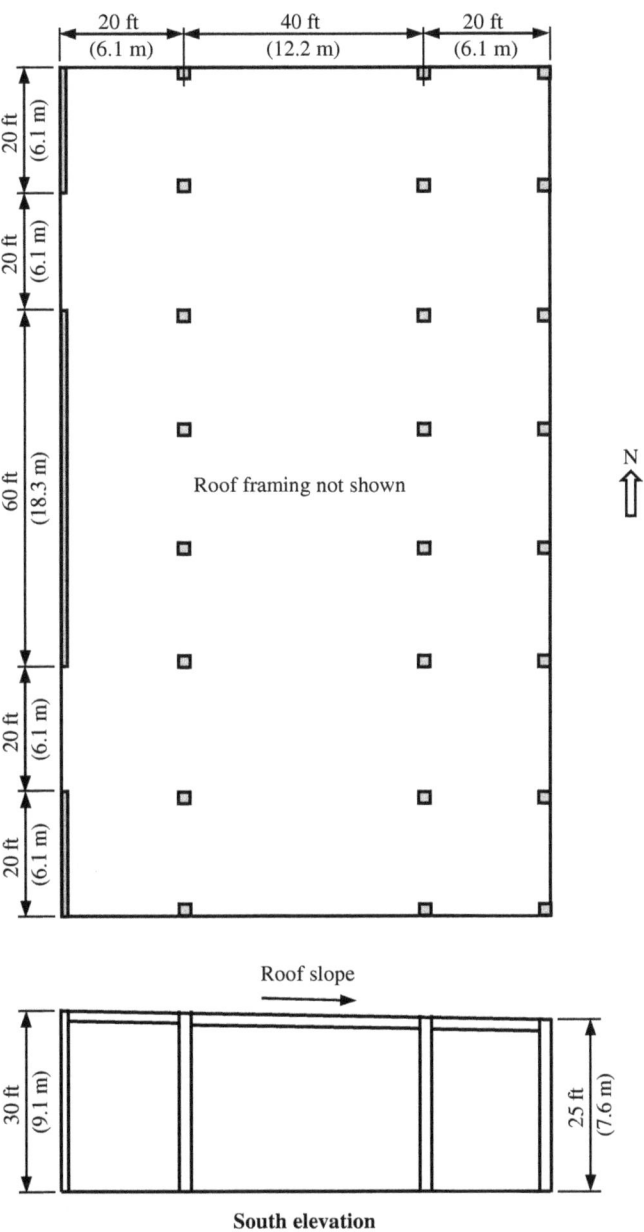

FIGURE 2.8   Material storage building in Example 2.4.

Building Surface	Opening Area, A_o, ft^2 (m^2)
North wall	$(80 \times 25) + (0.5 \times 80 \times 5) = 2,200$ (204.4)
South wall	$(80 \times 25) + (0.5 \times 80 \times 5) = 2,200$ (204.4)
East wall	$25.0 \times 140 = 3,500$ (325.2)
West wall	$2 \times 20 \times 15 = 600$ (55.7)
Roof	$2 \times 1 = 2$ (0.19)

TABLE 2.13   Opening Areas in the Material Storage Building in Example 2.4

by 15 ft (6.1 m by 4.6 m) openings. Two 1-ft^2 (0.093-m^2) openings are in the roof. A summary of the opening areas in each wall and the roof is given in Table 2.13.

Solution

Step 1—Determine if the building can be classified as open   Table 2.7

A building is defined as open where each wall is at least 80 percent open. A summary of wall opening percentages is given in Table 2.14. Because the west wall is less than 80 percent open, the building cannot be classified as open.

Step 2—Determine if the building can be classified as partially enclosed   Table 2.7

Check if the following condition is satisfied assuming the east wall is the windward wall:

$$A_o > 1.10 A_{oi}$$

For the east wall, $A_o = 3,500$ ft^2 (325.2 m^2)   Table 2.14

A_{oi} = sum of openings in the building envelope not including A_o

$$= 2,200 + 2,200 + 600 + 2 = 5,002 \text{ ft}^2$$

$$A_o = 3,500 \text{ ft}^2 < 1.10 A_{oi} = 1.10 \times 5,002 = 5,502 \text{ ft}^2$$

Building Surface	Opening Area, A_o, ft^2 (m^2)	Gross Area, A_g, ft^2 (m^2)	$(A_o/A_g) \times 100$ (%)
North wall	2,200 (204.4)	2,200 (204.4)	100
South wall	2,200 (204.4)	2,200 (204.4)	100
East wall	3,500 (325.2)	3,500 (325.2)	100
West wall	600 (55.7)	4,200 (390.2)	14.3

TABLE 2.14   Wall Opening Percentages for the Material Storage Building in Example 2.4

In S.I.:

$$A_{oi} = 204.4 + 204.4 + 55.7 + 0.19 = 464.7 \text{ m}^2$$

$$A_o = 325.2 \text{ m}^2 < 1.10 A_{oi} = 1.10 \times 464.7 = 511.2 \text{ m}^2$$

Therefore, the building cannot be classified as partially enclosed.

Step 3—Determine if the building can be classified as enclosed   *Table 2.7*

It is evident the building cannot be classified as enclosed because the condition that all the walls have openings with areas less than the smaller of 4 ft² (0.37 m²) and 1 percent the area of the wall is not satisfied.

Therefore, because this building does not comply with the conditions for open, partially enclosed, and enclosed buildings, the building is classified as partially open.

CHAPTER 3

Wind Loads on Buildings: MWFRS (Directional Procedure)

3.1 Overview

This chapter contains the requirements for determining wind pressures and loads on the main wind force resisting systems (MWFRSs) of enclosed, partially enclosed, and open buildings of all heights in accordance with the Directional Procedure of ASCE/SEI Chapter 27. Because the internal pressure coefficients for partially open buildings are the same as those for enclosed buildings, it is assumed the methods to determine wind loads on partially open buildings are the same as those applicable to enclosed buildings (see ASCE/SEI Table 26.13-1).

A summary of the wind load procedures in Chapter 27 is given in Table 3.1.

The provisions in Chapter 27 are applicable to buildings that comply with the following (ASCE/SEI 27.1.2):

- The building is regular-shaped, that is, the building has no unusual geometrical irregularities in spatial form.

- The building does not have response characteristics that make it subject to across-wind loading, vortex shedding, or instability caused by galloping or flutter. Additionally, the building is not located at a site where channeling effects or buffeting in the wake of upwind obstructions warrant special consideration.

Buildings not meeting these conditions must be designed by either recognized literature that documents such wind load effects or by the wind tunnel procedure in Chapter 31 (ASCE/SEI 27.1.3).

Reduction in wind pressure due to apparent shielding by surrounding buildings, other structures, or terrain features is not permitted (ASCE/SEI 27.1.4). Such shielding may be modified or completely removed during the lifespan of the building, which could result in significantly higher wind loads.

Minimum design wind pressures and loads are given in ASCE/SEI 27.1.5. For enclosed and partially enclosed buildings, the minimum wind loads to be used in the design of the MWFRS are equal to 16 lb/ft² (0.77 kN/m²) multiplied by the wall area of the building and 8 lb/ft² (0.38 kN/m²) multiplied by the roof area of the building projected onto a vertical plane normal to the wind direction. The minimum wall and roof wind loads are to be applied simultaneously. Application of minimum wind pressures are illustrated in Fig. 3.1 for wind along the two primary axes of a building. For open buildings, the minimum wind load is equal to 16 lb/ft² (0.77 kN/m²) multiplied by the area of the building either normal to the wind direction or projected on a plane normal to the wind direction. Minimum design pressures or loads are a separate load case from those specified in Part 1 or Part 2 of Chapter 27.

Part	Applicability		Conditions
	Building Type	**Height Limit**	
1	Enclosed	None	• Regular-shaped building
	Partially enclosed		• Building does not have response characteristics making it subject to across-wind loading, vortex shedding, instability due to galloping or flutter
	Open		• Building is not located at a site where channeling effects or buffeting in the wake of upwind obstructions warrant special consideration
2	Enclosed, simple diaphragm	$h \leq 160$ ft (48.8 m)	• Same conditions as in Part 1 • Building must meet the conditions for either a Class 1 or Class 2 building: Class 1 Building: 1. $h \leq 60$ ft (18.3 m) 2. $0.2 \leq L/B \leq 5.0$ Class 2 Building: 1. 60 ft (18.3 m) $< h \leq 160$ ft (48.8 m) 2. $0.5 \leq L/B \leq 2.0$ 3. $n_1 \geq 75/h$ (h in ft) [$n_1 \geq 22.86/h$ (h in m)]

TABLE 3.1   Wind Load Procedures in ASCE/SEI Chapter 27

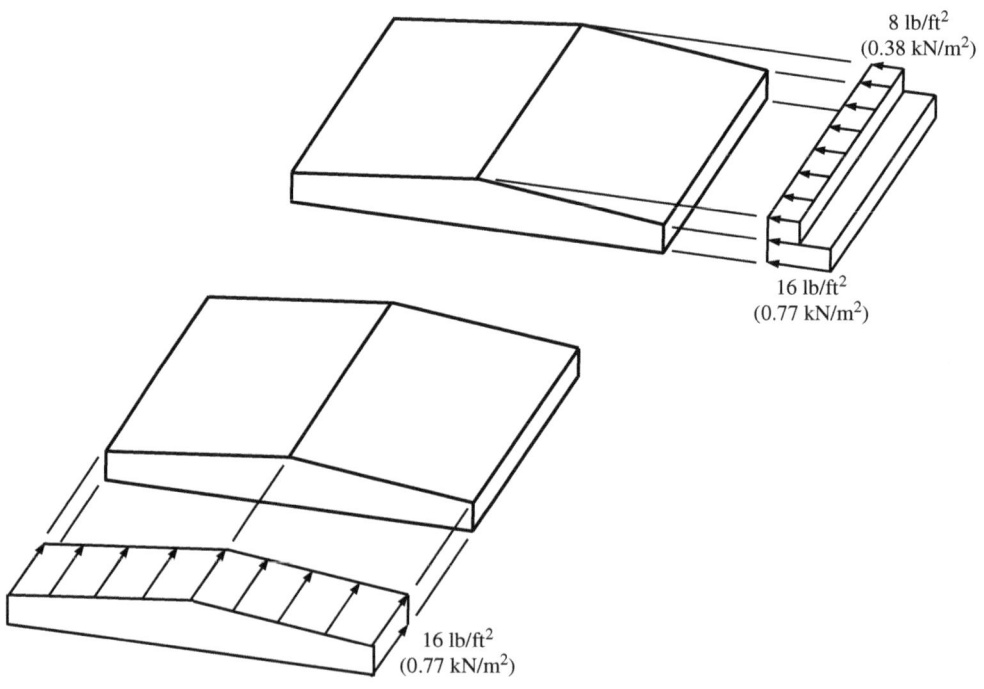

FIGURE 3.1   Application of minimum design wind pressures.

3.2 Enclosed, Partially Enclosed, and Open Buildings of All Heights (Part 1)

3.2.1 Enclosed and Partially Enclosed Rigid and Flexible Buildings

Design Wind Pressures, p

Design wind pressures, p, for the MWFRS of buildings of all heights are determined by ASCE/SEI Equation (27.3-1):

$$p = qGC_p - q_i(GC_{pi}) \tag{3.1}$$

This equation is used to calculate the wind pressures on the windward wall, leeward wall, side walls, and roof of a building. The first part of the equation corresponds to external pressures and the second part corresponds to internal pressures. The pressures are applied simultaneously to these surfaces, as shown in ASCE/SEI Figure 27.3-1 (see Fig. 3.2 for the application of external wind pressures on a building with a gable or hip roof).

Velocity pressure, q_z. The velocity pressure, q_z, is determined in accordance with ASCE/SEI 26.10 (see Sec. 2.7 of this publication). On the windward wall, this pressure varies with respect to height; on all other surfaces, it is calculated at the mean roof height, h (see Fig. 3.2).

Gust-effect factor, G. The gust-effect factor for rigid buildings, G, may be taken as 0.85 or can be calculated by ASCE/SEI Equation (26.11-6) (see Sec. 2.8 of this publication). For flexible buildings, G_f is determined by ASCE/SEI Equation (26.11-10) and is substituted for G in Eq. (3.1).

External pressure coefficients, C_p—Buildings with walls and flat, gable, hip, monoslope, or mansard roofs. External pressure coefficients, C_p, are given in ASCE/SEI Figure 27.3-1 for the wall and roof surfaces of a building. Wall pressure coefficients are constant on windward and side walls and vary with the aspect ratio of the building, L/B, on the leeward wall. Roof pressure coefficients vary with the ratio of the mean roof height to the plan dimension of the building (h/L) and with the roof angle (θ) for a given wind direction (normal to ridge or parallel to ridge). These pressure coefficients are intended to be used with q_h, and the parallel to ridge wind direction is applicable for flat roofs. Negative roof pressures increase as h/L increases. Also, as θ increases, negative pressure decreases until a roof angle is reached where the pressure becomes positive. Where two values of C_p are listed in the figure, the windward roof is subjected to either positive or negative pressure and the structure must be designed for both. Other important information on the use of ASCE/SEI Figure 27.3-1 is given in the notes below the tabulated pressure coefficients.

External pressure coefficients, C_p—Domed roofs with a circular base and arched roofs. External pressure coefficients for enclosed and partially enclosed domed roofs with a circular base and arched roofs are given in ASCE/SEI Figures 27.3-2 and 27.3-3, respectively.

Two load cases must be considered for domed roofs. In case A, pressure coefficients are determined between various locations on the dome by linear interpolation along arcs of the dome parallel to the direction of wind; this defines maximum uplift on the dome in many cases. In case B, the pressure coefficient is assumed to

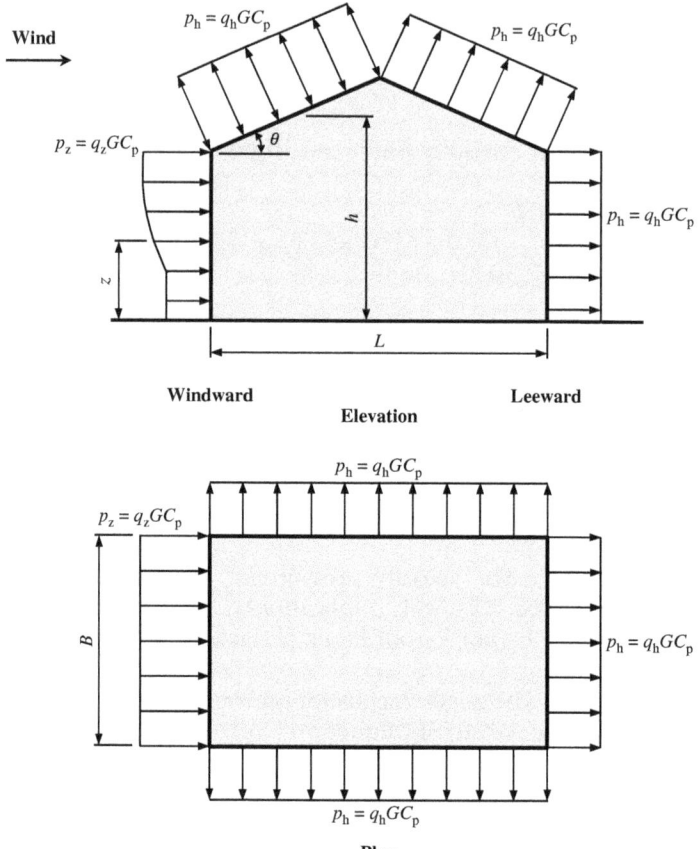

Surface	Pressure Coefficient, C_p		Velocity Pressure, q	Gust-Effect Factor	Wind Pressure, p
Windward wall	0.8		$q_z = 0.00256\,K_zK_{zt}K_dK_eV^2$ (lb/ft^2) $q_z = 0.613\,K_zK_{zt}K_dK_eV^2$ (N/m^2)		$p_z = q_zGC_p$
Leeward wall	L/B	C_p	$q_h = 0.00256\,K_hK_{zt}K_dK_eV^2$ (lb/ft^2) $q_h = 0.613\,K_hK_{zt}K_dK_eV^2$ (N/m^2)	• For rigid buildings, use G • For flexible buildings, use G_f	$p_h = q_hGC_p$
	0 – 1	–0.5			
	2	–0.3			
	≥ 4	–0.2			
Side walls	–0.7				
Windward roof Leeward roof	See ASCE/SEI Figure 27.3-1				

FIGURE 3.2 Application of external wind pressures on building with a gable or hip roof (Directional Procedure, Part 1).

be a constant value at a specific point on the dome for angles less than or equal to 25 degrees and is determined by linear interpolation from 25 degrees to other points on the dome. Wind tunnel tests are recommended for domes that are larger than 200 ft (61.0 m) in diameter and in cases where resonant response can be an issue (see ASCE/SEI C27.4.1).

Velocity pressure for internal pressure determination, q_i. The velocity pressure for internal pressure determination, q_i, is determined as follows:

- On all the surfaces of enclosed buildings and for negative internal pressure evaluation in partially enclosed buildings: $q_i = q_h$.
- For positive internal pressure evaluation in partially enclosed buildings: $q_i = q_z$ where height z is defined as the level of the highest opening in the building that could affect the positive internal pressure.

It is conservative to set q_i equal to q_h in all cases where positive internal pressure is evaluated.

For buildings located in windborne debris regions, glazing that is not impact-resistant or protected with an impact-resistant covering must be treated as an opening, and q_i is to be determined accordingly.

Internal pressure coefficients (GC_{pi}). Internal pressure coefficients, (GC_{pi}), are determined in accordance with ASCE/SEI 26.13 (see Sec. 2.10 of this publication). Both positive and negative values of (GC_{pi}) must be considered in order to establish the critical load effects.

The effects from internal pressure cancel out when evaluating the total horizontal wind pressure on the MWFRS of a building. Thus, the total horizontal pressure at any height z above ground in the direction of wind is equal to the external pressure p_z on the windward face at height z plus the external pressure p_h on the leeward face (see Fig. 3.2).

Design Procedure

A step-by-step procedure to determine the design wind pressures on enclosed and partially enclosed rigid and flexible buildings with walls and flat, gable, hip, monoslope, or mansard roofs is given in Fig. 3.3.

3.2.2 Open Buildings with Monoslope, Pitched, or Troughed Free Roofs

Design Wind Pressures, p

Net design pressure, p, for the MWFRS of open buildings with monoslope, pitched, or troughed free roofs is determined by ASCE/SEI Equation (27.3-2):

$$p = q_h G C_N \qquad (3.2)$$

In this equation, q_h is the velocity pressure at the mean roof height and G is the gust-effect factor determined in accordance with ASCE/SEI 26.11.

Net pressure coefficients C_N are given in ASCE/SEI Figures 27.3-4 through 27.3-7 for various roof configurations. Load cases A and B are identified in the figures. Both load cases must be considered in order to obtain the maximum load effects for a particular roof slope and blockage configuration.

The magnitude of roof pressure in open buildings is highly dependent on the blockage configuration beneath the roof. Goods or materials stored under the roof can restrict air flow, which can introduce significant upward pressures on the bottom surface of the roof. The net pressure coefficients for clear wind flow are to be used in cases where blockage is less than or equal to 50 percent. Obstructed wind flow is applicable where the blockage is greater than 50 percent. In cases where the usage below the roof is not evident, both unobstructed and obstructed load cases should be investigated.

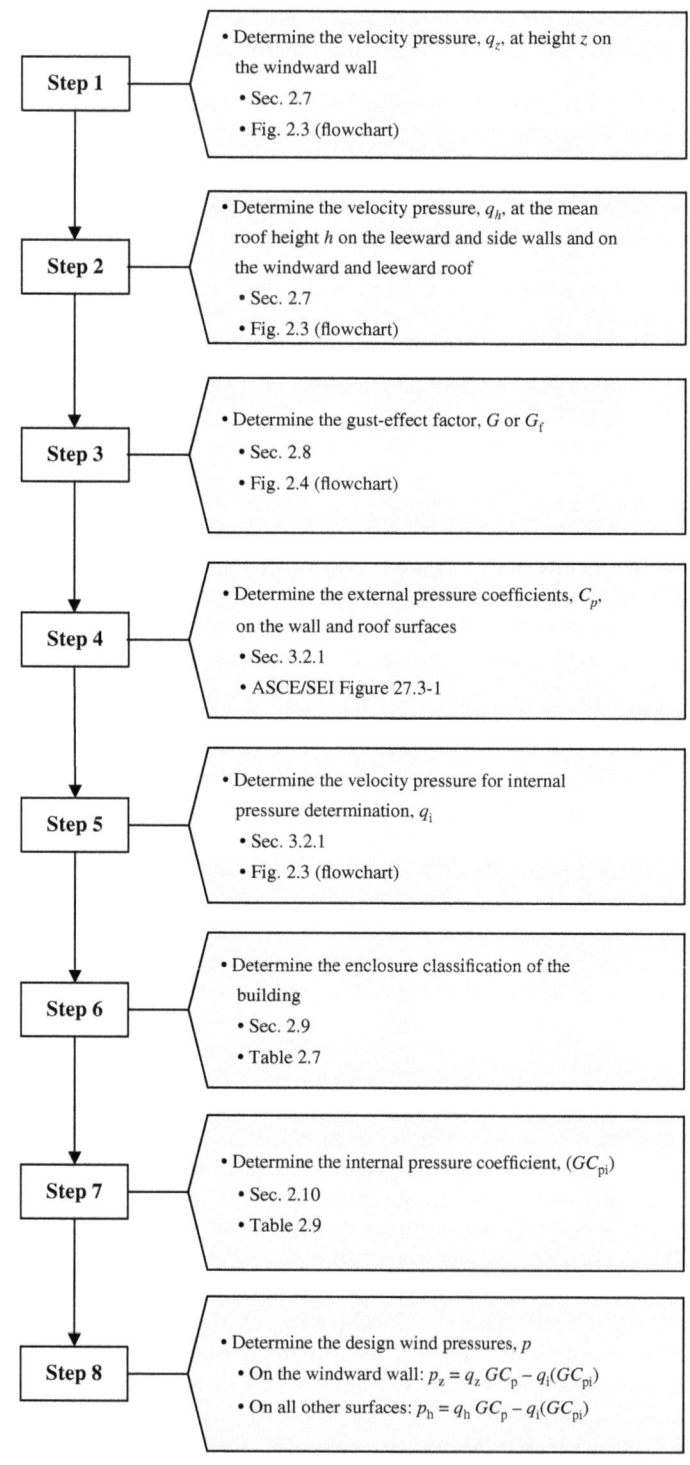

Step 1
- Determine the velocity pressure, q_z, at height z on the windward wall
- Sec. 2.7
- Fig. 2.3 (flowchart)

Step 2
- Determine the velocity pressure, q_h, at the mean roof height h on the leeward and side walls and on the windward and leeward roof
- Sec. 2.7
- Fig. 2.3 (flowchart)

Step 3
- Determine the gust-effect factor, G or G_f
- Sec. 2.8
- Fig. 2.4 (flowchart)

Step 4
- Determine the external pressure coefficients, C_p, on the wall and roof surfaces
- Sec. 3.2.1
- ASCE/SEI Figure 27.3-1

Step 5
- Determine the velocity pressure for internal pressure determination, q_i
- Sec. 3.2.1
- Fig. 2.3 (flowchart)

Step 6
- Determine the enclosure classification of the building
- Sec. 2.9
- Table 2.7

Step 7
- Determine the internal pressure coefficient, (GC_{pi})
- Sec. 2.10
- Table 2.9

Step 8
- Determine the design wind pressures, p
- On the windward wall: $p_z = q_z\, GC_p - q_i(GC_{pi})$
- On all other surfaces: $p_h = q_h\, GC_p - q_i(GC_{pi})$

FIGURE 3.3   Procedure to determine design wind pressures, p, on enclosed and partially enclosed rigid and flexible buildings with walls and flat, gable, hip, monoslope, or mansard roofs (Directional Procedure, Part 1).

Additional Wind Pressures

For structures with free roofs containing fascia panels where the angle of the plane of the roof is less than or equal to 5 degrees, the fascia panels must be considered an inverted parapet (ASCE/SEI 27.3.2). The contribution of the wind loads on the fascia to the wind loads on the MWFRS is to be determined using the requirements in ASCE/SEI 27.3.4 with q_p in ASCE/SEI Equation (27.3-3) taken as q_h.

An additional horizontal wind load must be applied to open or partially enclosed buildings with transverse frames and pitched roofs where the angle of the plane of the roof from the horizontal is less than or equal to 45 degrees. This force is to be determined using the provisions in ASCE/SEI 28.3.5 and must act in combination with the roof loads calculated in accordance with ASCE/SEI 27.3.3. As shown in ASCE/Figure 28.3-2, this horizontal force is applied to the building parallel to the ridge of the roof.

Design Procedure

A step-by-step procedure to determine the design wind pressures on open buildings with monoslope, pitched, or troughed free roofs is given in Fig. 3.4.

3.2.3   Roof Overhangs

The positive external pressure on the bottom surface of a windward roof overhang is equal to the following (ASCE/SEI 27.3.3):

$$p_z = q_z GC_p \qquad (3.3)$$

where p_z and q_z are evaluated at the top of the windward wall located a distance z above ground level and the pressure coefficient C_p is equal to the pressure coefficient for the windward wall (that is, $C_p = 0.8$; see ASCE/SEI 27.3.3).

The positive pressure on the bottom surface of the overhang is combined with the wind pressure on the top surface determined in accordance with ASCE/SEI Figure 27.3-1 (see Fig. 3.5).

Provisions are not provided for wind pressures on the bottom surface of a leeward overhang.

3.2.4   Parapets

Design wind pressures for the effects of parapets, p_p, on the MWFRS of rigid or flexible buildings with flat, gable, or hip roofs are determined by ASCE/SEI Equation (27.3-3):

$$p_p = q_p(GC_{pn}) \qquad (3.4)$$

In this equation, q_p is the velocity pressure evaluated at the top of the parapet using Eqs. (2.4) and (2.5) of this publication and (GC_{pn}) is the combined net pressure coefficient, which is equal to +1.5 for a windward parapet and –1.0 for a leeward parapet.

As shown in Fig. 3.6, a windward parapet experiences positive wall pressure on the exterior side (front surface) and negative roof pressure on the roof side (back surface). The behavior on the back surface is based on the assumption that the zone of negative pressure caused by the wind flow separation at the eave of the roof moves up to the top of the parapet, resulting in the back side of the parapet having the same negative pressure as the roof.

A leeward parapet experiences a positive wall pressure on the roof side (back surface) and a negative wall pressure on the exterior side (front surface). It is assumed that

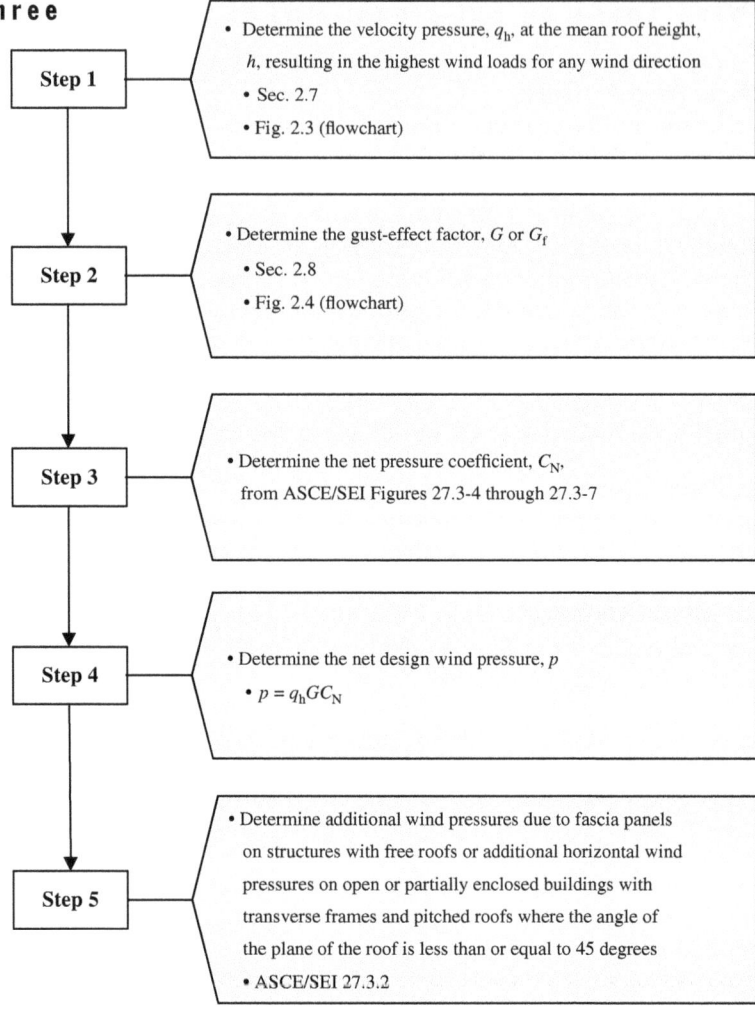

Step 1
- Determine the velocity pressure, q_h, at the mean roof height, h, resulting in the highest wind loads for any wind direction
- Sec. 2.7
- Fig. 2.3 (flowchart)

Step 2
- Determine the gust-effect factor, G or G_f
- Sec. 2.8
- Fig. 2.4 (flowchart)

Step 3
- Determine the net pressure coefficient, C_N, from ASCE/SEI Figures 27.3-4 through 27.3-7

Step 4
- Determine the net design wind pressure, p
- $p = q_h G C_N$

Step 5
- Determine additional wind pressures due to fascia panels on structures with free roofs or additional horizontal wind pressures on open or partially enclosed buildings with transverse frames and pitched roofs where the angle of the plane of the roof is less than or equal to 45 degrees
- ASCE/SEI 27.3.2

Figure 3.4 Procedure to determine design wind pressures, p, on open buildings with monoslope, pitched, or troughed free roofs (Directional Procedure, Part 1).

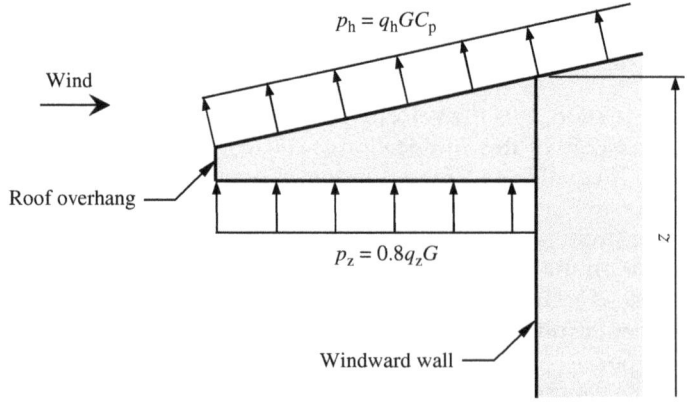

$p_h = q_h G C_p$

Wind

Roof overhang

$p_z = 0.8 q_z G$

Windward wall

Figure 3.5 Wind pressures on a roof overhang (Directional Procedure, Part 1).

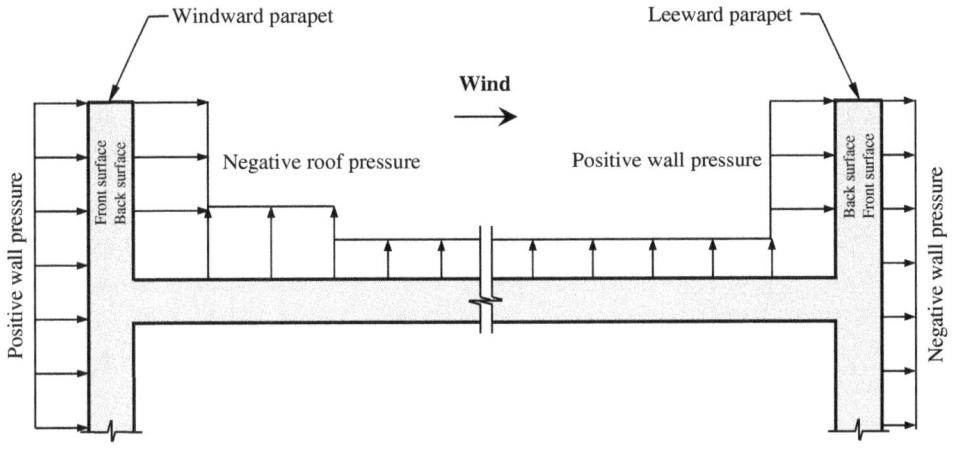

Figure 3.6   Wind pressures on parapets.

the windward and leeward parapets are separated a sufficient distance so that shielding by the windward parapet does not decrease the positive wall pressure on the leeward parapet.

For simplicity, the pressure p_p is the combined net pressure due to the combination of the net pressures from the front and back surfaces of the parapet, which is captured by the combined net pressure coefficients (GC_{pn}) for windward and leeward parapets (see Fig. 3.7). Because wind can occur in any direction, a parapet must be designed for both sets of pressures. The internal pressures inside the parapet cancel out in the determination of the combined pressure coefficient.

The pressures determined on the parapets are combined with the external pressures on the building to obtain the total wind pressures on the MWFRS.

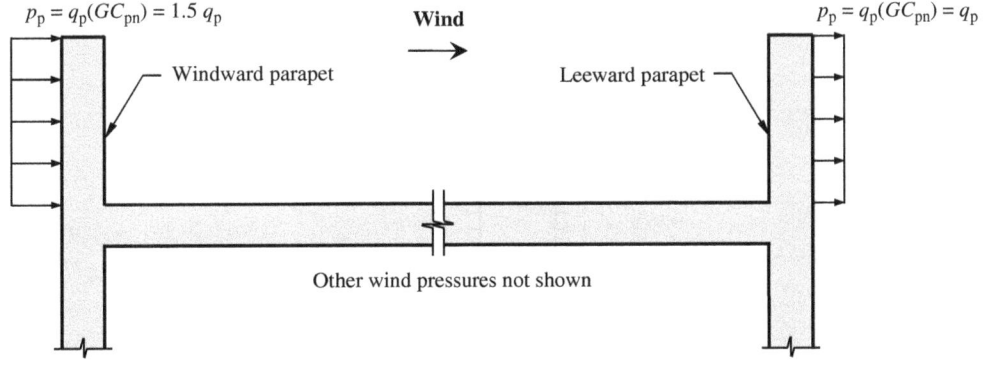

Figure 3.7   Design wind pressures on parapets (Directional Procedure, Part 1).

3.2.5   Design Wind Load Cases

The MWFRS of buildings of all heights subjected to the wind pressures determined by Part 1 of the Directional Procedure must be designed for the load cases depicted in ASCE/SEI Figure 27.3-8 (see Fig. 3.8; in this figure, plan views of the building are shown, the subscripts x and y refer to the principal axes of the building, and W and L refer to the windward and leeward faces, respectively).

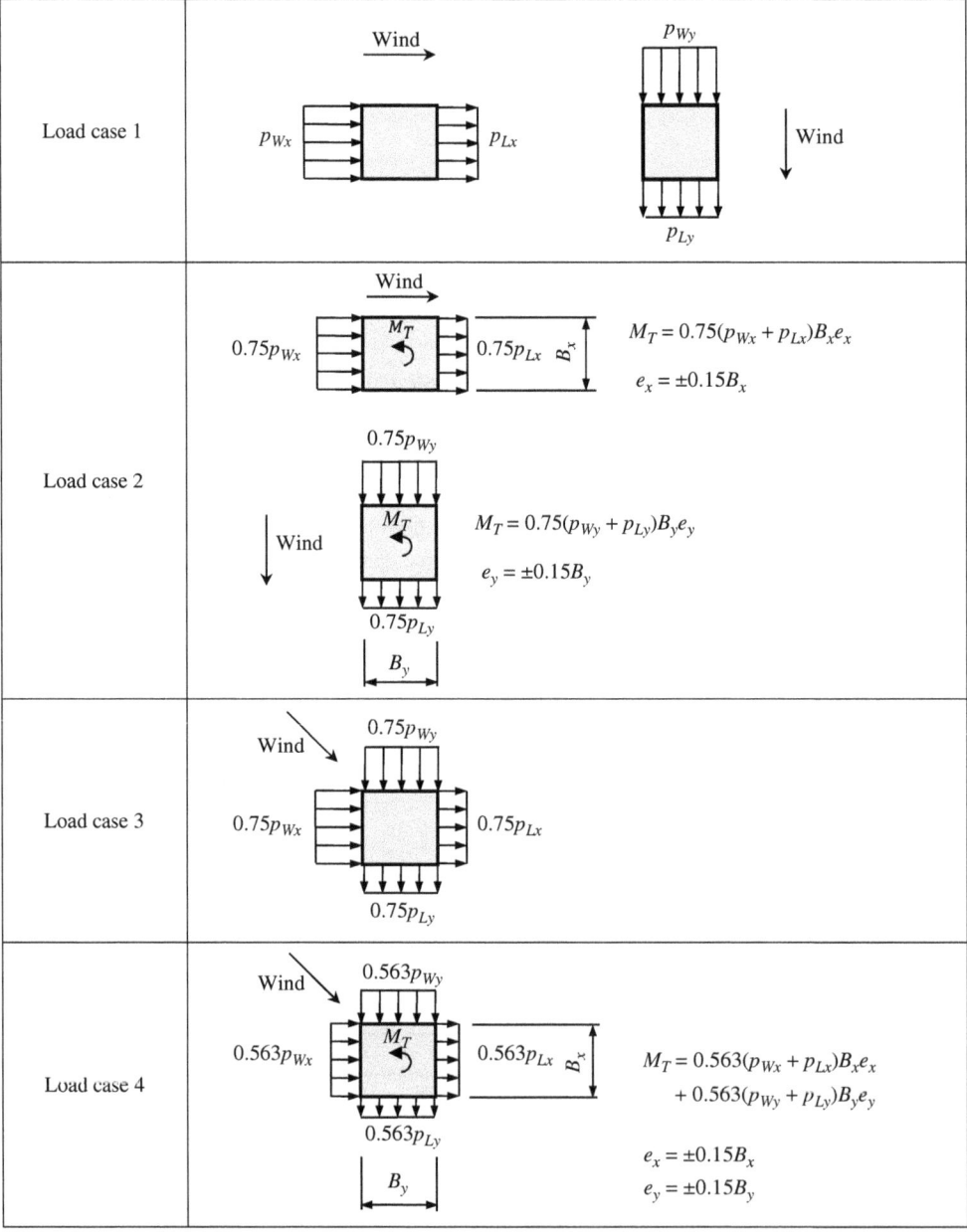

FIGURE 3.8   Design wind load cases on the MWFRS (Directional Procedure).

In load case 1, design wind pressures are applied along the principal axes of a building separately.

Load case 2 accounts for the effects of nonuniform pressure on different faces of the building due to wind flow. Nonuniform pressures introduce torsion on the building, and this is accounted for in design by subjecting the building to 75 percent of the design wind pressures applied along the principal axis of the building plus a torsional moment M_T determined using an eccentricity equal to 15 percent of the appropriate plan dimension of the building. Torsional effects are determined in each principal direction separately.

A critical load case can occur when the design wind load acts diagonally to a building. This is accounted for in load case 3, where 75 percent of the maximum design wind pressures are applied along the principal axes of a building simultaneously.

Load case 4 considers the effects due to diagonal wind loads and torsion. Seventy-five percent of the wind pressures in load case 2 are applied along the principal axes of a building simultaneously, and a torsional moment is applied, which is determined using 15 percent of the plan dimensions of the building.

The exception in ASCE/SEI 27.3.5 permits buildings that meet the requirements of Section D.1 of ASCE/SEI Appendix D to be designed for load cases 1 and 3 only. The following buildings do not need to be designed for the effects from torsion:

- One-story buildings with a mean roof height of less than or equal to 30 ft (9.2 m).

- Buildings two stories or less framed with light-frame construction.

- Buildings two stories or less with flexible diaphragms.

- Buildings controlled by seismic loading (see ASCE/SEI D.3.1 and D.3.2 for criteria for buildings with diaphragms that are not flexible and those that are flexible, respectively).

- Buildings classified as torsionally regular under wind loads (ASCE/SEI D.4).

- Buildings with flexible diaphragms and designed for at least 1.5 times the design wind pressures in cases 1 and 3.

- Class 1 and Class 2 simple diaphragm buildings with a mean roof height less than or equal to 160 ft (48.8 m) meeting the requirements of ASCE/SEI D.6 (see Fig. 3.9 of this publication for the definition of Class 1 and Class 2 simple diaphragm buildings).

In the case of flexible buildings, dynamic effects can increase the effects from torsion. ASCE/SEI Equation (27.3-4) accounts for these effects. The eccentricity e determined by this equation is to be used in the appropriate load cases in Fig. 3.8 in lieu of the eccentricities e_x and e_y given in that figure for rigid structures. An eccentricity must be considered for each principal axis of the building, and the sign of the eccentricity must be plus or minus, whichever causes the more severe load case.

3.3 Enclosed Simple Diaphragm Buildings with $h \leq 160$ ft (48.8 m) (Part 2)

3.3.1 Overview

Simplified methods are given in Part 2 of Chapter 27 for determining wind pressures for the MWFRS of enclosed, simple diaphragm buildings with mean roof heights less than or equal to 160 ft (48.8 m) that meet the additional conditions in ASCE/SEI 27.4.2 (see Table 3.1 of this publication).

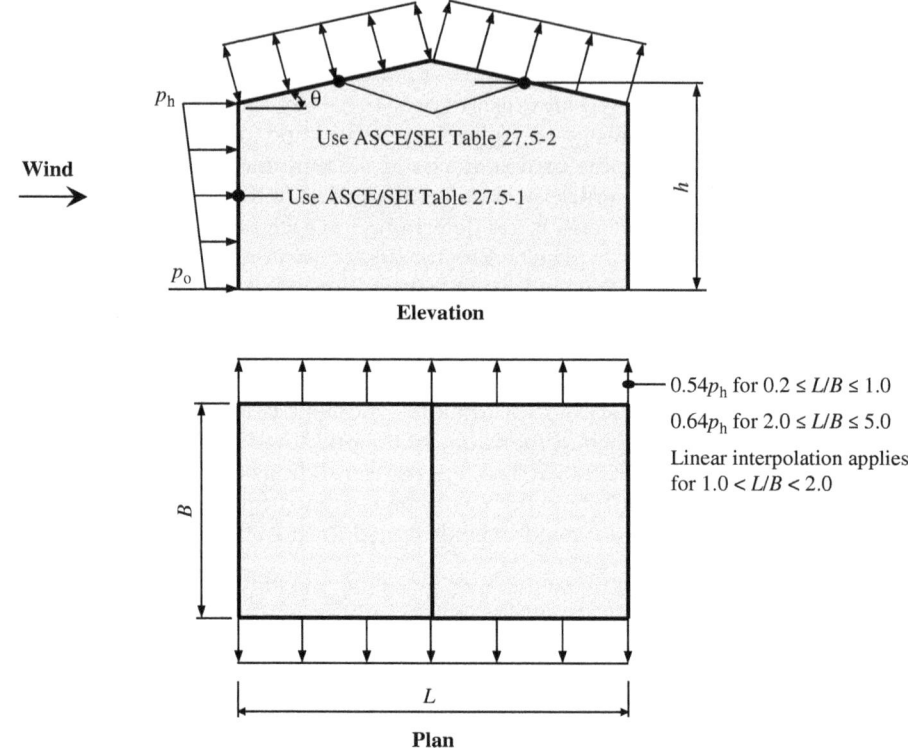

Wind

Elevation

Use ASCE/SEI Table 27.5-2

Use ASCE/SEI Table 27.5-1

$0.54p_h$ for $0.2 \le L/B \le 1.0$

$0.64p_h$ for $2.0 \le L/B \le 5.0$

Linear interpolation applies for $1.0 < L/B < 2.0$

Plan

Building Class	Conditions
1	• $h \le 60$ ft (18.3 m) • $0.2 \le L/B \le 5.0$
2	• 60 ft (18.3 m) $< h \le$ 160 ft (48.8 m) • $0.5 \le L/B \le 2.0$ • $n_1 \ge 75/h$ (h in ft)[$n_1 \ge 22.86/h$ (h in m)]

Figure 3.9   Wind pressures on the MWFRS (Directional Procedure, Part 2).

A simple diaphragm building is one in which both windward and leeward wind loads are transmitted by roof and vertically spanning wall assemblies through continuous floor and roof diaphragms to the MWFRS (ASCE/SEI 26.2). As such, internal wind pressures cancel out in the determination of the total wind load in the direction of analysis. Buildings with structural expansion joints in diaphragms or structural systems with girts or other horizontal members that transfer significant wind loads directly to vertical members of the MWFRS are not permitted to be designed using the methods in Part 2.

The wind pressures determined by ASCE/SEI 27.5 must be multiplied by the topographic factor, K_{zt}, which is calculated at $0.33h$ (see ASCE/SEI 27.4.4 and Sec. 2.5 of this publication). Alternatively, it is permitted to determine the wind pressures in

ASCE/SEI Tables 27.5-1 and 27.5-2 with a wind velocity equal to $V\sqrt{K_{zt}}$ where K_{zt} is determined at $0.33h$.

The design procedures in Part 2 apply to buildings with either rigid or flexible diaphragms. Diaphragms constructed of untopped metal deck, concrete-filled metal deck, and concrete slabs with a span-to-depth ratio of 2 or less are permitted to be idealized as rigid (ASCE/SEI 27.4.5). Diaphragms constructed of wood panels are permitted to be idealized as flexible.

3.3.2 Design Wind Pressures on the MWFRS

Wall and Roof Surfaces

Net design wind pressures for the walls and roof surfaces of Classes 1 and 2 buildings can be determined directly from ASCE/SEI Tables 27.5-1 and 27.5-2, respectively. Wall pressures, which have been calculated using the procedures in Part 1, are tabulated for exposures B, C, and D as a function of wind velocity V, mean roof height h, and building aspect ratio L/B. The top pressure in the table is defined as p_h and the bottom pressure is defined as p_o (see Fig. 3.9). Interpolation between these values is permitted (see Note 5 in ASCE/SEI Table 27.5-1). For Class 1 buildings with an aspect ratio $L/B < 0.5$, the tabulated wind pressures for $L/B = 0.5$ are to be used (ASCE/SEI 27.5.1). Similarly, for Class 1 buildings with $L/B > 2.0$, the tabulated wind pressures for $L/B = 2.0$ are to be used.

Along-wind net wind pressures are distributed over the height of the building as shown in ASCE/SEI Figure 27.5-1 and Fig. 3.9 and are applied to the projected area of the building walls in the direction of the wind. Exterior side wall pressures are constant and are applied to the projected area of the building normal to the direction of the wind.

Tabulated roof pressures are given in ASCE/SEI Table 27.5-2 for Exposure C as a function of V, h, and roof slope. For exposures B and D, the tabulated roof pressures are to be multiplied by the appropriate exposure adjustment factors in that figure. Roof pressures are applied perpendicular to the roof surfaces, as shown in ASCE/SEI Figure 27.5-1 and Fig. 3.9. The different zones over which these pressures are to be distributed are identified in ASCE/SEI Table 27.5-2 for flat, gable, hip, monoslope, and mansard roofs. Roof pressures are given for two load cases and both must be investigated where applicable. Load case 2 is required when investigating maximum overturning effects on the building due to the wind pressures.

Pressures on the walls and the roof must be applied simultaneously to the building, as shown in ASCE/SEI Figure 27.5-1 and Fig. 3.9 (ASCE/SEI 27.5.1). Also, the MWFRS must be designed for the load cases defined in ASCE/SEI Figure 27.3-8 (see Fig. 3.8). The torsional load cases in ASCE/SEI Figure 27.3-8 (cases 2 and 4) need not be considered for buildings meeting the requirements of ASCE/SEI Appendix D.

A step-by-step procedure to determine the design wind pressures on the walls and roof of an enclosed, simple diaphragm building with a mean roof height less than or equal to 160 ft (48.8 m) in accordance with ASCE/SEI 27.5 is given in Fig. 3.10.

The minimum wind pressures in ASCE/SEI 27.1.5 must also be considered (see Fig. 3.1).

Parapets

The additional pressure on the MWFRS due to roof parapets is equal to 2.25 times the wall pressure from ASCE/SEI Table 27.5-1 using an aspect ratio of $L/B = 1.0$ and a height h_p, which is equal to the distance from the ground to the top of the parapet (ASCE/SEI 27.5.2). This net horizontal pressure accounts for the windward and

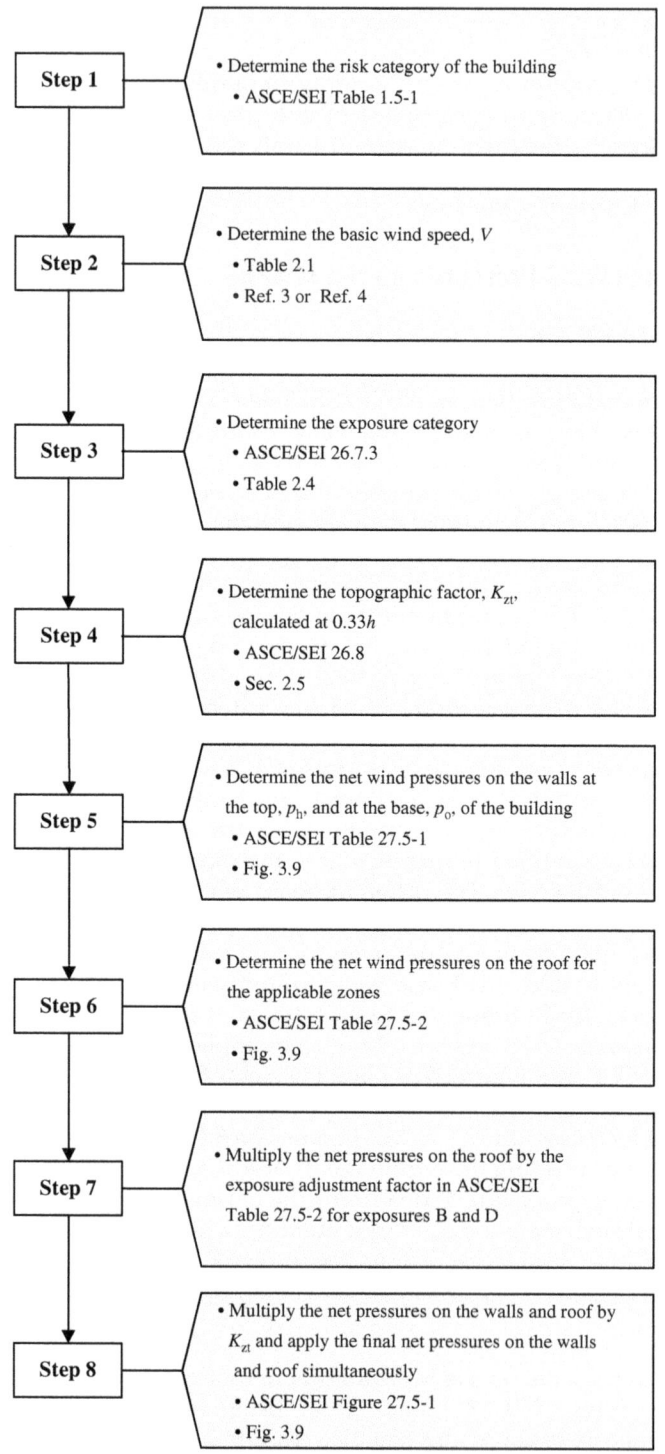

Figure 3.10 Procedure to determine design wind pressures on an enclosed, simple diaphragm building with $h \leq 160$ ft (48.8 m) (Directional Procedure, Part 2).

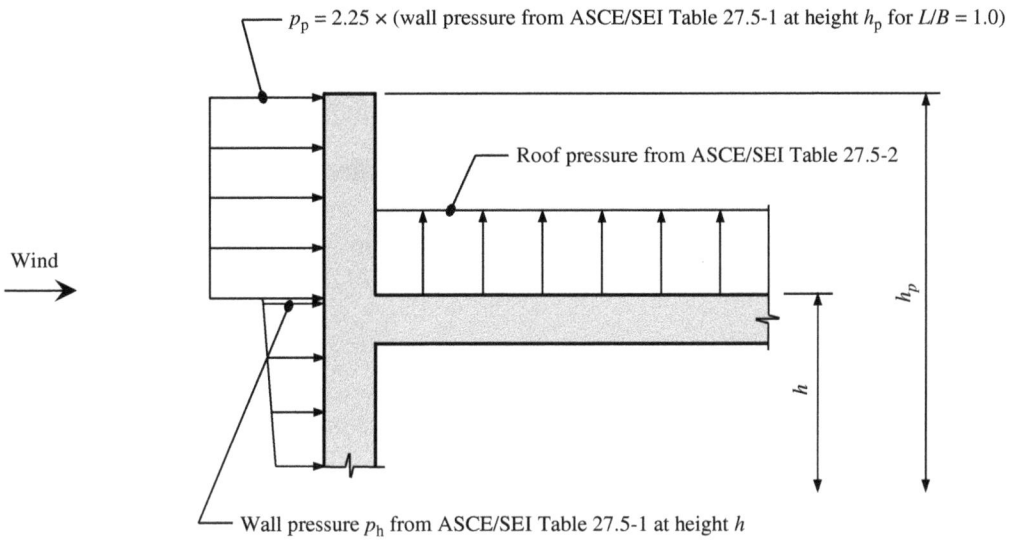

$p_p = 2.25 \times$ (wall pressure from ASCE/SEI Table 27.5-1 at height h_p for $L/B = 1.0$)

Roof pressure from ASCE/SEI Table 27.5-2

Wind

h_p

h

Wall pressure p_h from ASCE/SEI Table 27.5-1 at height h

Figure 3.11 Net wind pressure on parapets (Directional Procedure, Part 2).

leeward parapet loading on both the windward and leeward building surfaces, and is applied to the projected area of the parapet surface simultaneously with the net wall and roof pressures (see ASCE/SEI Figure 27.5-2 and Fig. 3.11).

Roof Overhangs

The effect of vertical wind loads on roof overhangs is given in ASCE/SEI 27.5.3. A positive wind pressure equal to 75 percent of the roof edge pressure from ASCE/SEI Table 27.5-2 for Zone 1 or Zone 3, whichever is applicable, must be applied to the underside of the windward overhang (see ASCE/SEI Fig. 27.5-3 and Fig. 3.12).

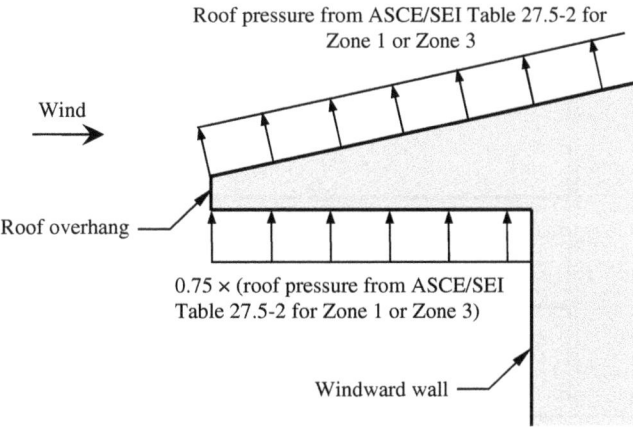

Roof pressure from ASCE/SEI Table 27.5-2 for Zone 1 or Zone 3

Wind

Roof overhang

$0.75 \times$ (roof pressure from ASCE/SEI Table 27.5-2 for Zone 1 or Zone 3)

Windward wall

Figure 3.12 Wind pressure on roof overhangs (Directional Procedure, Part 2).

3.4 Examples

The following examples illustrate the determination of wind pressures using parts 1 and 2 of the Directional Procedure.

3.4.1 Example 3.1—Wind Pressures on a Commercial Building, MWFRS, Chapter 27, Part 1

Determine the wind pressures in both directions on the MWFRS of the commercial building in Fig. 3.13 using the requirements in Part 1 of Chapter 27 and the design data in Table 3.2.

Solution

Check if the building meets all the conditions in ASCE/SEI 27.1.2 so that Part 1 in Chapter 27 can be used to determine the design wind pressures on the MWFRS.

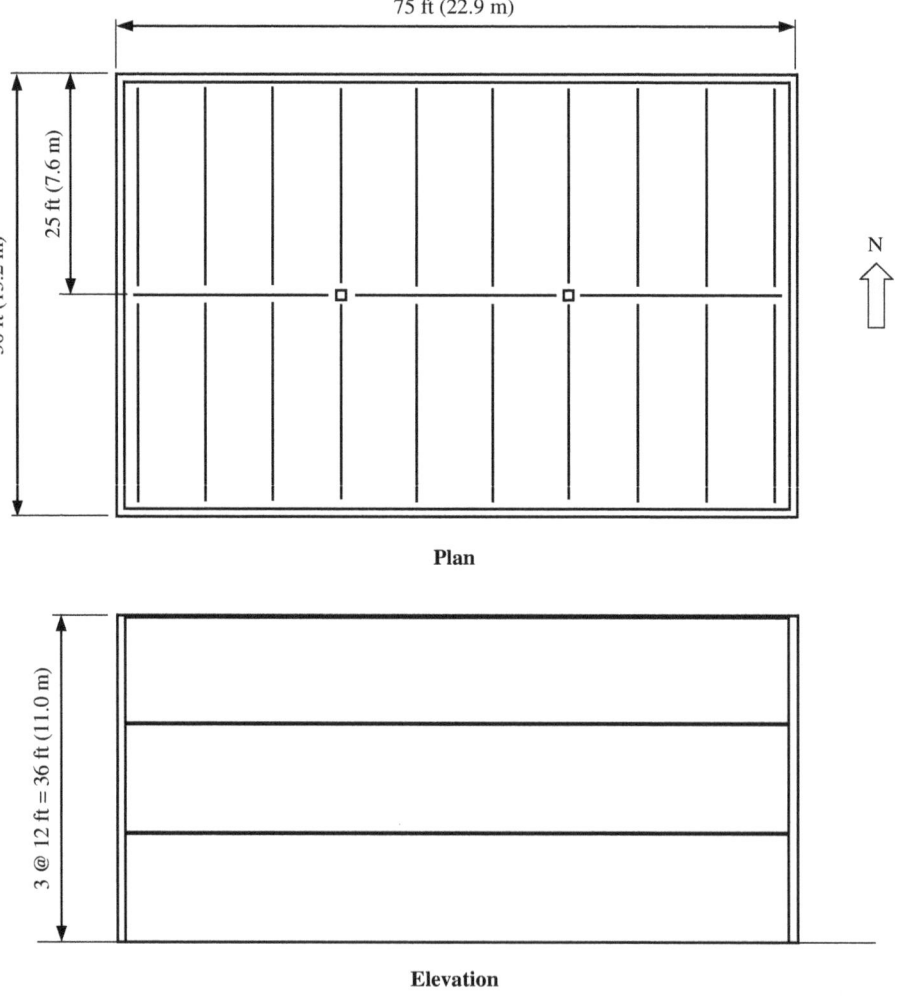

75 ft (22.9 m)

25 ft (7.6 m)

50 ft (15.2 m)

N

Plan

3 @ 12 ft = 36 ft (11.0 m)

Elevation

FIGURE 3.13 Plan and elevation of the commercial building in Example 3.1.

Location	Phoenix, AZ
Surface roughness	C
Topography	Not situated on a hill, ridge, or escarpment
Occupancy	Less than 300 people congregate in one area at the same time
Enclosure classification	Enclosed
MWFRS	7-in. (178-mm) thick precast concrete walls at the perimeter of the building

TABLE 3.2   Design Data for the Commercial Building in Example 3.1

The building is regular-shaped and does not have any unusual geometric irregularities in spatial form. Also, the building does not have any response characteristics that make it subject to across-wind loading or similar effects, and it is not sited at a location where channeling effects or buffeting in the wake of upwind obstructions need to be considered.

Therefore, the conditions in ASCE/SEI 27.1.2 and 27.1.3 are met and the requirements in Part 1 of Chapter 27 may be used.

The design procedure in Fig. 3.3 is used to determine the design wind pressures, p, in both directions.

Step 1—Determine the velocity pressure, q_z, over the height of the windward wall   Fig. 2.3

The flowchart in Fig. 2.3 is used to determine q_z.

- Step 1a—Determine the surface roughness category   ASCE/SEI 26.7.2

  In Table 3.2, the surface roughness is given as C.

- Step 1b—Determine the exposure category   ASCE/SEI 26.7.3

  It is assumed that surface roughness C applies in all directions and that exposures B and D are not applicable. Therefore, the exposure category is C.

- Step 1c—Determine the terrain exposure constants   ASCE/SEI Table 26.11-1

  For Exposure C, $\alpha = 9.5$ and $z_g = 900$ ft (274.32 m).

- Step 1d—Determine the velocity pressure exposure coefficient, K_z

  ASCE/SEI Table 26.10-1

  Values of K_z are given in Table 3.3 for Exposure C.

  For example, at $z = 36$ ft (11.0 m):

$$K_z = 2.01(z/z_g)^{2/\alpha} = 2.01 \times (36/900)^{2/9.5} = 1.02$$

  In S.I.: $K_z = 2.01 \times (11.0/274.32)^{2/9.5} = 1.02$

- Step 1e—Determine the topographic factor, K_{zt} ASCE/SEI 26.8

  Because the building is not located on a hill, ridge, or escarpment, $K_{zt} = 1.0$.

- Step 1f—Determine the wind directionality factor, K_d ASCE/SEI Table 26.6-1

  For the MWFRS of a building structure, $K_d = 0.85$.

- Step 1g—Determine the ground elevation factor, K_e ASCE/SEI 26.9

  Ground elevation factor can be taken as 1.0 for all elevations.

Height above Ground Level, z, ft (m)	K_z
36 (11.0)	1.02
24 (7.3)	0.94
12 (3.7)	0.85

TABLE 3.3   Velocity Pressure Exposure Coefficient, K_z, for the Building in Example 3.1

- Step 1h—Determine the risk category of the building   ASCE/SEI Table 1.5-1

  Due to the nature of its occupancy, this commercial building falls under Risk Category II.

- Step 1i—Determine the basic wind speed, V Table 2.1

  For Risk Category II, use IBC Figure 1609.3(1) or ASCE/SEI Figure 26.5-1B.

  Equivalently, use Ref. 3 or Ref. 4 to obtain $V = 101$ mi/h (45 m/s) for Phoenix, AZ.

- Step 1j—Determine the wind velocity pressure, q_z ASCE/SEI 26.10.2

$$q_z = 0.00256 K_z K_{zt} K_d K_e V^2$$

$$= 0.00256 \times K_z \times 1.0 \times 0.85 \times 1.0 \times 101^2 = 22.2 K_z \text{ lb/ft}^2$$

In S.I.:

$$q_z = 0.613 K_z K_{zt} K_d K_e V^2$$

$$= 0.613 \times K_z \times 1.0 \times 0.85 \times 1.0 \times 45^2 / 1,000 = 1.06 K_z \text{ kN/m}^2$$

The velocity pressures over the height of the windward wall are given in Table 3.4.

Step 2—Determine the velocity pressure, q_h, on the roof and side walls   *Fig. 2.3*

From Step 1, $q_h = 22.6$ lb/ft^2 $(1.08$ kN/m$^2)$.

Step 3—Determine the gust-effect factor   *Fig. 2.4*

- Step 3a—Determine the fundamental natural frequency, n_1

  In lieu of obtaining n_1 from a dynamic analysis of the building, check if the approximate lower bound natural frequency, n_a, given in ASCE/SEI 26.11.3 can be used for this building:

- Building height = 36 ft (11.0 m) < 300 ft (91.4 m)

Height above Ground Level, z, ft (m)	K_z	q_z, lb/ft^2 (kN/m^2)
36 (11.0)	1.02	22.6 (1.08)
24 (7.3)	0.94	20.9 (1.00)
12 (3.7)	0.85	18.9 (0.90)

TABLE 3.4   Velocity Pressure, q_z, for the Building in Example 3.1

- Building height = 36 ft (11.0 m) < $4L_{eff} = 4 \times 50 = 200$ ft (61.0 m) in the N-S direction
  = 36 ft (11.0 m) < $4L_{eff} = 4 \times 75 = 300$ ft (91.4 m) in the E-W direction

Because both of these limitations are satisfied, ASCE/SEI Equation (26.11-5) can be used to determine n_a in both directions for the MWFRS consisting of precast concrete walls.

In the N-S direction:

A_b = base area of the building = $50 \times 75 = 3{,}750$ ft^2 (348.4 m^2)

$h = h_i$ = height of walls = 36 ft (11.0 m)

A_i = horizontal cross-sectional area of wall = $(7/12) \times 50 = 29.2$ ft^2 (2.7 m^2)

D_i = length of wall = 50 ft (15.2 m)

$$C_w = \frac{100}{A_B} \sum_{i=1}^{n} \left(\frac{h}{h_i}\right)^2 \frac{A_i}{\left[1 + 0.83\left(\frac{h_i}{D_i}\right)^2\right]} = \frac{2 \times 100}{3{,}750} \times \left(\frac{36}{36}\right)^2 \times \frac{29.2}{\left[1 + 0.83\left(\frac{36}{50}\right)^2\right]} = 1.09$$

$n_a = 385(C_w)^{0.5}/h = 385 \times (1.09)^{0.5}/36 = 11.2$ Hz > 1 Hz

In S.I.:

$$C_w = \frac{100}{A_B} \sum_{i=1}^{n} \left(\frac{h}{h_i}\right)^2 \frac{A_i}{\left[1 + 0.83\left(\frac{h_i}{D_i}\right)^2\right]} = \frac{2 \times 100}{348.4} \times \left(\frac{11.0}{11.0}\right)^2 \times \frac{2.7}{\left[1 + 0.83\left(\frac{11.0}{15.2}\right)^2\right]} = 1.08$$

$n_a = 117.3(C_w)^{0.5}/h = 117.3 \times (1.08)^{0.5}/11 = 11.1$ Hz > 1 Hz

Therefore, the building is rigid in the N-S direction.

Similar calculations in the E-W direction result in $n_a > 1$ Hz, so the building is rigid in that direction as well.

- Step 3b—Determine the gust-effect factor, G

For rigid buildings, G is permitted to be taken as 0.85.   ASCE/SEI 26.11.4

Step 4—Determine the external pressure coefficients, C_p   *ASCE/SEI Fig. 27.3-1*

For a flat roof, assume a ridge line occurs parallel to the long dimension of the building. In this example, the ridge line is along the interior column line in the east-west direction.

- N-S wind

Windward wall: $C_p = 0.8$

Leeward wall: $L/B = 50/75 = 0.67$; $C_p = -0.5$

Side walls: $C_p = -0.7$

Roof: normal to the ridge with $\theta < 10$ degrees and $h/L = 36/50 = 0.72$

From 0 to 18 ft (0 to 5.5 m): $C_p = -0.9 + \dfrac{(-1.3 + 0.9) \times (-0.72 + 0.5)}{-1.0 + 0.5} = -1.08, -0.18$

(the value of −1.3 was not reduced linearly with the area over which it is applicable in accordance with footnote b in ASCE/SEI Figure 27.3-1)

From 18 ft to 36 ft (5.5 m to 11.0 m):

$$C_p = -0.7 + \frac{(-0.9 + 0.7) \times (-1.0 + 0.72)}{-1.0 + 0.5} = -0.81, -0.18$$

From 36 ft to 50 ft (11.0 m to 15.2 m):

$$C_p = -0.5 + \frac{(-0.7 + 0.5) \times (-0.72 + 0.5)}{-1.0 + 0.5} = -0.59, -0.18$$

- E-W wind

  Windward wall: $C_p = 0.8$

  Leeward wall: $L/B = 75/50 = 1.5$; $C_p = \dfrac{-0.5 - 0.3}{2} = -0.4$

  Side walls: $C_p = -0.7$

  Roof: parallel to ridge with $h/L = 36/75 = 0.48$

  From 0 to 18 ft (0 to 5.5 m): $C_p = -0.9, -0.18$

  From 18 ft to 36 ft (5.5 m to 11.0 m): $C_p = -0.9, -0.18$

  From 36 ft to 72 ft (11.0 m to 22.0 m): $C_p = -0.5, -0.18$

  From 72 ft to 75 ft (22.0 m to 22.9 m): $C_p = -0.3, -0.18$

The roof pressure coefficient $C_p = -0.18$ may become critical where wind loads are combined with roof live loads or snow loads. Determination of wind pressures based on this pressure coefficient should be performed, but such calculations are not shown in this example.

Step 5—Determine the velocity pressure for internal pressure determination, q_i Sec. 3.2.1

According to ASCE/SEI 27.3.1, $q_i = q_h = 22.6$ lb/ft^2 (1.08 kN/m^2)

Step 6—Determine the enclosure classification   Table 2.7

In the design data, the building is given as enclosed.

Step 7—Determine the internal pressure coefficient, GC_{pi} Table 2.9

For an enclosed building, $GC_{pi} = +0.18, -0.18$.

Step 8—Determine the design wind pressures, p

- Windward walls for wind in both directions

$$p_z = q_z GC_p - q_h(GC_{pi})$$
$$= (q_z \times 0.85 \times 0.8) - [22.6 \times (\pm 0.18)]$$
$$= (0.68q_z \mp 4.1) \text{ lb/ft}^2$$

In S.I.:

$$p_z = (q_z \times 0.85 \times 0.8) - [1.08 \times (\pm 0.18)] = (0.68 q_z \mp 0.19) \text{ kN/m}^2$$

- Leeward wall, side walls, and roof

$$\begin{aligned} p_h &= q_h GC_p - q_h(GC_{pi}) \\ &= (22.6 \times 0.85 \times C_p) - [22.6 \times (\pm 0.18)] \\ &= (19.21 C_p \mp 4.1) \text{ lb/ft}^2 \end{aligned}$$

In S.I.:

$$p_h = (1.08 \times 0.85 \times C_p) - [1.08 \times (\pm 0.18)] = (0.92 C_p \mp 0.19) \text{ kN/m}^2$$

Design wind pressures for wind in the N-S and E-W directions are given in Tables 3.5 and 3.6, respectively.

Net design wind pressures in the N-S and E-W directions for positive and negative internal pressures are given in Figs. 3.14 and 3.15, respectively.

Building Surface	Height above Ground Level, z, ft (m)	q, lb/ft² (kN/m²)	External Pressure qGC_p, lb/ft² (kN/m²)	Internal Pressure q_h(GC_pi), lb/ft² (kN/m²)	Net Pressure, p, lb/ft² (kN/m²)	
					(+GC_pi)	(−GC_pi)
Windward wall	36 (11.0)	22.6 (1.08)	15.4 (0.73)	±4.1 (±0.19)	11.3 (0.54)	19.5 (0.92)
	24 (7.3)	20.9 (1.00)	14.2 (0.68)	±4.1 (±0.19)	10.1 (0.49)	18.3 (0.87)
	12 (3.7)	18.9 (0.90)	12.9 (0.61)	±4.1 (±0.19)	8.8 (0.42)	17.0 (0.80)
Leeward wall	All	22.6 (1.08)	−9.6 (−0.46)	±4.1 (±0.19)	−13.7 (−0.65)	−5.5 (−0.27)
Side walls	All	22.6 (1.08)	−13.5 (−0.64)	±4.1 (±0.19)	−17.6 (−0.83)	−9.4 (−0.45)
Roof	___(1)	22.6 (1.08)	−20.8 (−0.99)	±4.1 (±0.19)	−24.9 (−1.18)	−16.7 (−0.80)
	___(2)	22.6 (1.08)	−15.6 (−0.74)	±4.1 (±0.19)	−19.7 (−0.93)	−11.5 (−0.55)
	___(3)	22.6 (1.08)	−11.3 (−0.54)	±4.1 (±0.19)	−15.4 (−0.73)	−7.2 (−0.35)

(1) From windward edge of roof to 18 ft (5.5 m).
(2) From 18 ft (5.5 m) to 36 ft (11.0 m).
(3) From 36 ft (11.0 m) to 50 ft (15.2 m).

TABLE 3.5 Design Wind Pressures, p, for Wind in the N-S Direction for the Building in Example 3.1

Building Surface	Height above Ground Level, z, ft (m)	q, lb/ft² (kN/m²)	External Pressure qGC_p, lb/ft² (kN/m²)	Internal Pressure $q_h(GC_{pi})$, lb/ft² (kN/m²)	Net Pressure, p, lb/ft² (kN/m²)	
					(+GC_{pi})	(−GC_{pi})
Windward wall	36 (11.0)	22.6 (1.08)	15.4 (0.73)	±4.1 (±0.19)	11.3 (0.54)	19.5 (0.92)
	24 (7.3)	20.9 (1.00)	14.2 (0.68)	±4.1 (±0.19)	10.1 (0.49)	18.3 (0.87)
	12 (3.7)	18.9 (0.90)	12.9 (0.61)	±4.1 (±0.19)	8.8 (0.42)	17.0 (0.80)
Leeward wall	All	22.6 (1.08)	−7.7 (−0.37)	±4.1 (±0.19)	−11.8 (−0.56)	−3.6 (−0.18)
Side walls	All	22.6 (1.08)	−13.5 (−0.64)	±4.1 (±0.19)	−17.6 (−0.83)	−9.4 (−0.45)
Roof	—(1)	22.6 (1.08)	−17.3 (−0.83)	±4.1 (±0.19)	−21.4 (−1.02)	−13.2 (−0.64)
	—(2)	22.6 (1.08)	−17.3 (−0.83)	±4.1 (±0.19)	−21.4 (−1.02)	−13.2 (−0.64)
	—(3)	22.6 (1.08)	−9.6 (−0.46)	±4.1 (±0.19)	−13.7 (−0.65)	−5.5 (−0.27)
	—(4)	22.6 (1.08)	−5.8 (−0.28)	±4.1 (±0.19)	−9.9 (−0.47)	−1.7 (−0.09)

(1) From windward edge of roof to 18 ft (5.5 m).
(2) From 18 ft (5.5 m) to 36 ft (11.0 m).
(3) From 36 ft (11.0 m) to 72 ft (22.0 m).
(4) From 72 ft (22.0 m) to 75 ft (22.9 m).

TABLE 3.6   Design Wind Pressures, p, for Wind in the E-W Direction for the Building in Example 3.1

The horizontal wind pressures on the walls for both directions are shown in Fig. 3.16. It is evident from Figs. 3.14, 3.15, and 3.16 that the internal pressures cancel out when determining the horizontal wind pressures on the walls.

The building must be designed for the wind load cases defined in ASCE/SEI Figure 27.3-8 (see Fig. 3.8).

In case 1, the full design wind pressures act on the projected area perpendicular to each principal axis of the building at each level above ground. These pressures act separately along each principal axis. The windward and leeward pressures in Fig. 3.16 fall under case 1.

According to the exception in ASCE/SEI 27.3.5, buildings that meet the requirements of ASCE/SEI Appendix D.1 need only be designed for cases 1 and 3. It can be shown that this building is classified as a torsionally regular building under wind loads, which satisfies the requirements in ASCE/SEI D.4. Therefore, only cases 1 and 3 must be considered.

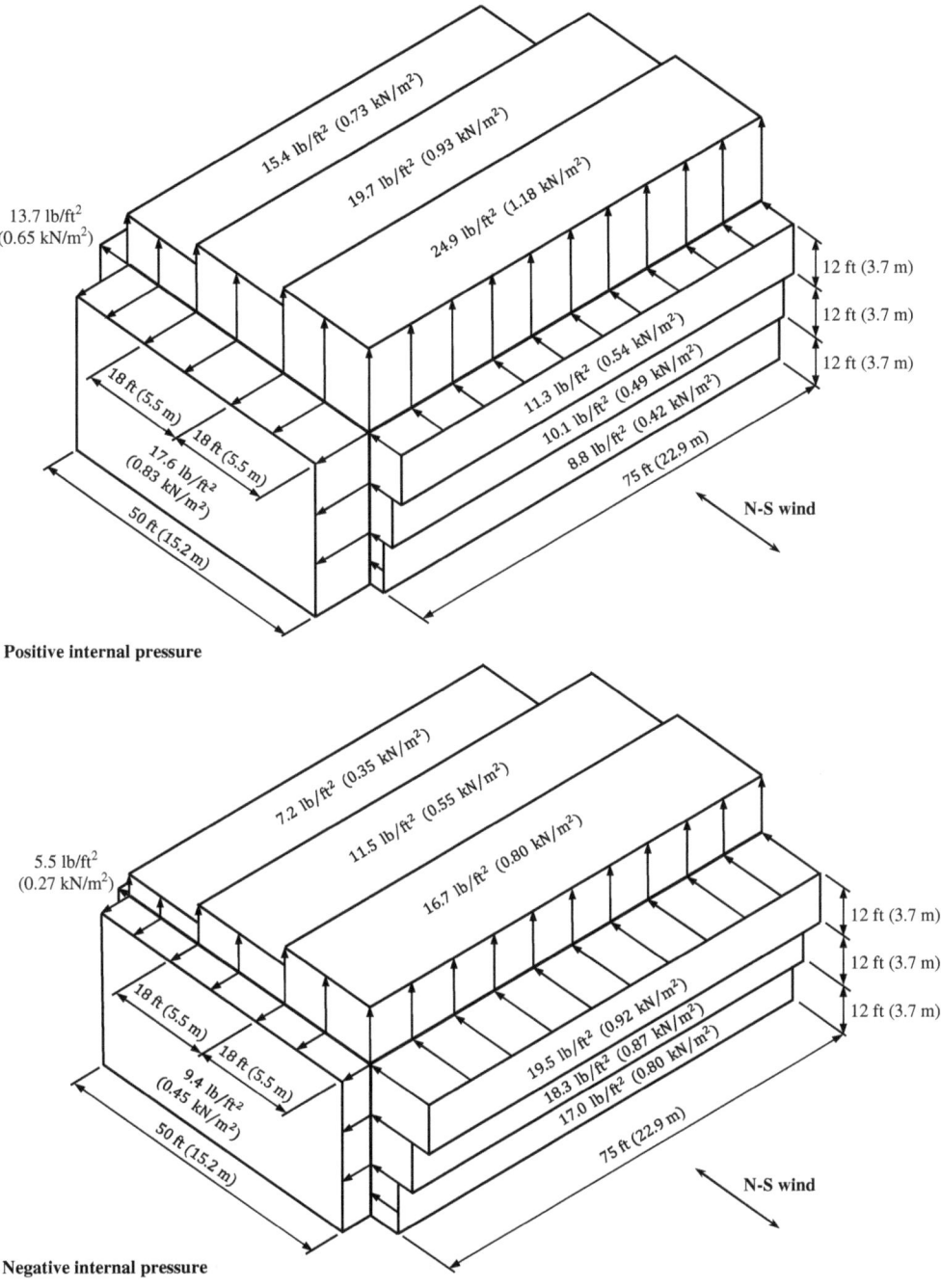

Positive internal pressure

Negative internal pressure

FIGURE 3.14 Net design wind pressures in the N-S direction for the building in Example 3.1.

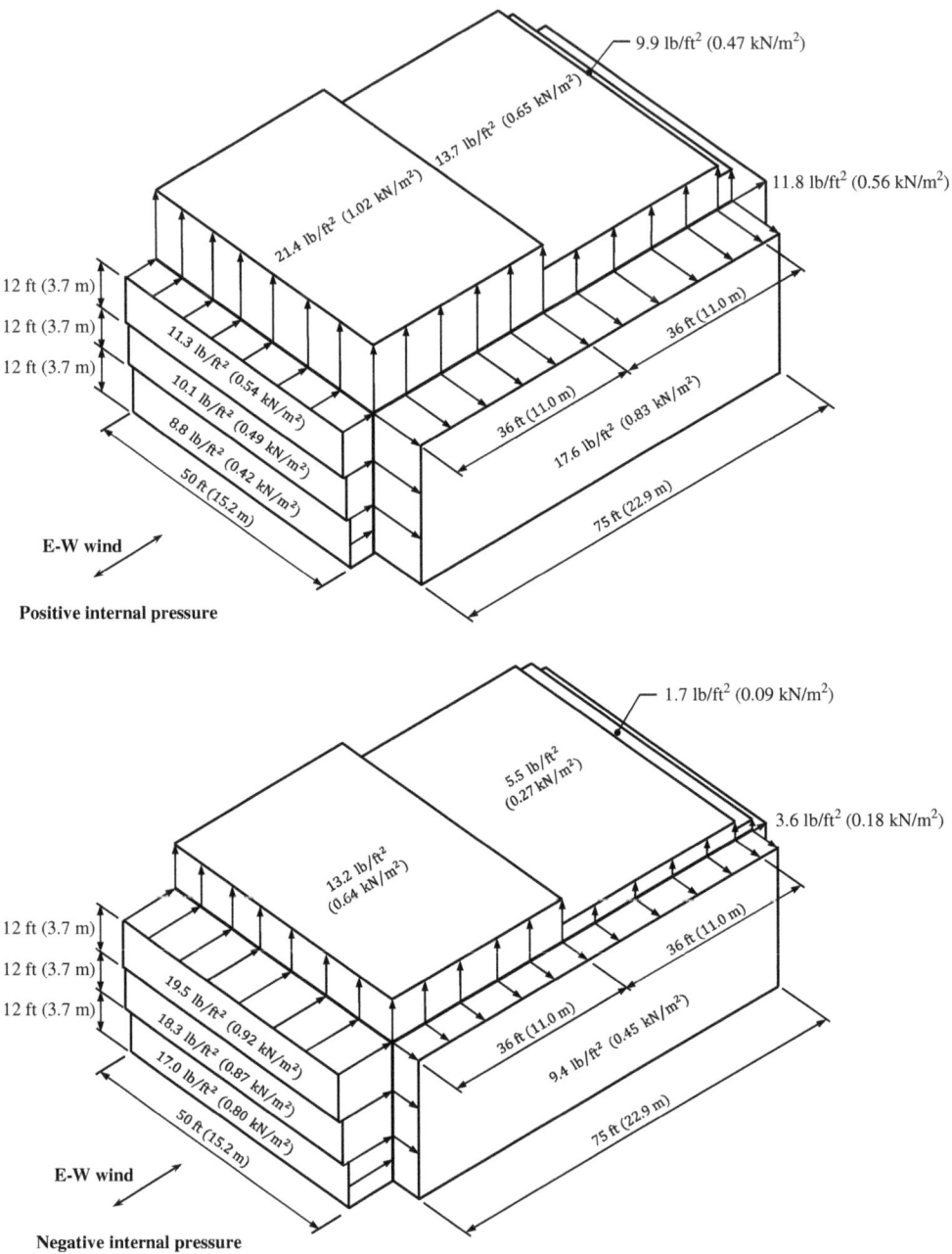

9.9 lb/ft² (0.47 kN/m²)

13.7 lb/ft² (0.65 kN/m²)

11.8 lb/ft² (0.56 kN/m²)

21.4 lb/ft² (1.02 kN/m²)

12 ft (3.7 m)

12 ft (3.7 m)

12 ft (3.7 m)

11.3 lb/ft² (0.54 kN/m²)

10.1 lb/ft² (0.49 kN/m²)

8.8 lb/ft² (0.42 kN/m²)

50 ft (15.2 m)

36 ft (11.0 m)

36 ft (11.0 m)

17.6 lb/ft² (0.83 kN/m²)

75 ft (22.9 m)

E-W wind

Positive internal pressure

1.7 lb/ft² (0.09 kN/m²)

5.5 lb/ft² (0.27 kN/m²)

3.6 lb/ft² (0.18 kN/m²)

13.2 lb/ft² (0.64 kN/m²)

12 ft (3.7 m)

12 ft (3.7 m)

12 ft (3.7 m)

19.5 lb/ft² (0.92 kN/m²)

18.3 lb/ft² (0.87 kN/m²)

17.0 lb/ft² (0.80 kN/m²)

50 ft (15.2 m)

36 ft (11.0 m)

36 ft (11.0 m)

9.4 lb/ft² (0.45 kN/m²)

75 ft (22.9 m)

E-W wind

Negative internal pressure

FIGURE 3.15   Net design wind pressures in the E-W direction for the building in Example 3.1.

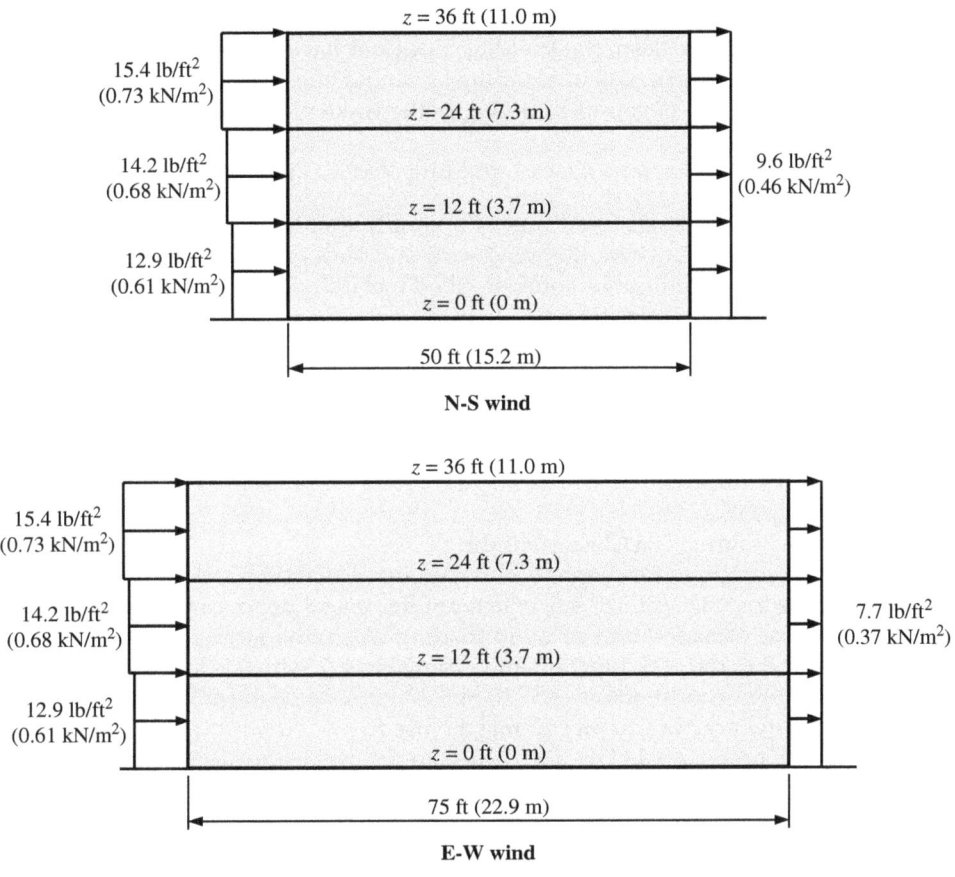

FIGURE 3.16   Horizontal wind pressures on the walls for the building in Example 3.1.

In case 3, 75 percent of the windward and leeward wall pressures in Fig. 3.16 act simultaneously on the building at each level above ground.

The minimum design wind loading prescribed in ASCE/SEI 27.1.5 must be considered as a load case in addition to the load cases above.

3.4.2   Example 3.2—Wind Pressures on a Commercial Building, MWFRS, Chapter 27, Part 2

Determine the wind pressures in both directions on the MWFRS of the commercial building in Fig. 3.13 using the requirements in Part 2 of Chapter 27 and the design data in Table 3.2. Assume the floor and roof diaphragms are concrete-filled metal decks, which transfer the lateral load effects to the precast concrete walls.

Solution
Check if the building meets all the conditions in ASCE/SEI 27.1.2, 27.4.2, and 27.4.5 so that Part 2 in Chapter 27 can be used to determine the design wind pressures on the MWFRS.

The building is regular-shaped and does not have any unusual geometric irregularities in spatial form. Also, the building does not have any response characteristics that make it subject to across-wind loading or similar effects, and it is not sited at a location where channeling effects or buffeting in the wake of upwind obstructions need to be considered.

Check the conditions for a Class 1 building (ASCE/SEI 27.4.2):

1. The building is enclosed and is a simple diaphragm building as defined in ASCE/SEI 26.2 where the windward and leeward wind loads are transmitted through the continuous concrete-filled metal decks (diaphragms) to the precast concrete walls (MWFRS).

2. Mean roof height $h = 36$ ft (11.0 m) < 60 ft (18.3 m).

3. In the N-S direction, $L/B = 50/75 = 0.67$ (in S.I.: 15.2/22.9 = 0.66), which is greater than 0.2 and less than 5.0.

In the E-W direction, $L/B = 75/50 = 1.5$ (in S.I.: 22.9/15.2 = 1.5), which is greater than 0.2 and less than 5.0.

Therefore, the building is a Class 1 building.

Check the condition for diaphragm flexibility (ASCE/SEI 27.4.5):

According to ASCE/SEI 27.4.5, concrete-filled metal decks can be considered rigid diaphragms for consideration of wind loading. Also, the largest span-to-depth ratio occurs for wind in the N-S direction and is equal to 1.5, which is less than 2.0.

Therefore, the conditions in ASCE/SEI 27.1.2, 27.4.2, and 27.4.5 are met and the requirements in Part 2 of Chapter 27 may be used.

The design procedure in Fig. 3.10 is used to determine the design wind pressures, p, in both directions.

Step 1—Determine the risk category of the building   *ASCE/SEI Table 1.5-1*

Due to the nature of its occupancy, this commercial building falls under Risk Category II.

Step 2—Determine the basic wind speed, V   *Table 2.1*

For Risk Category II, use IBC Figure 1609.3(1) or ASCE/SEI Figure 26.5-1B.

Equivalently, use Ref. 3 or Ref. 4 to obtain $V = 101$ mi/h (45 m/s) for Phoenix, AZ.

Step 3—Determine the exposure category   *ASCE/SEI 26.7.3*

It is assumed that surface roughness C applies in all directions and that exposures B and D are not applicable. Therefore, the exposure category is C.

Step 4—Determine the topographic factor, K_{zt}, calculated at 0.33h   *ASCE/SEI 26.8*

Because the building is not located on a hill, ridge, or escarpment, $K_{zt} = 1.0$.

Step 5—Determine the net wind pressures p_h and p_o on the walls   *ASCE/SEI Table 27.5-1*

- N-S wind

  The along-wind net wall pressures are obtained by reading the values from ASCE/SEI Table 27.5-1 for Exposure C, a wind velocity of 110 mi/h (49 m/s) [the pressures are subsequently adjusted for a wind velocity of 101 mi/h (45 m/s)], a mean roof height of 36 ft (11.0 m), and $L/B = 50/75 = 0.67$ (in S.I.: 15.2/22.9 = 0.66).

At the top of the wall: $p_h = 29.9$ lb/ft² (1.43 kN/m²) (by linear interpolation)

At the bottom of the wall: $p_o = 27.4$ lb/ft² (1.31 kN/m²) (by linear interpolation)

These pressures are multiplied by $(101)^2/(110)^2 = 0.84$ [In S.I.: $(45)^2/(49)^2 = 0.84$] to obtain the net pressures for $V = 101$ mi/h (45 m/s):

At the top of the wall:

$$p_h = 0.84 \times 29.9 = 25.1 \text{ lb/ft}^2 \text{ (in S.I.}: 0.84 \times 1.43 = 1.20 \text{ kN/m}^2)$$

At the bottom of the wall:

$$p_o = 0.84 \times 27.4 = 23.0 \text{ lb/ft}^2 \text{ (in S.I.}: 0.84 \times 1.31 = 1.10 \text{ kN/m}^2)$$

According to note 2 in ASCE/SEI Table 27.5-1, the uniform side wall external pressures are equal to $0.54p_h = 0.54 \times (-25.1) = -13.6$ lb/ft² [in S.I.: $0.54 \times (-1.20) = -0.65$ kN/m²].

- E-W wind

  The along-wind net wall pressures are obtained by reading the values from ASCE/SEI Table 27.5-1 for Exposure C, a wind velocity of 110 mi/h (49 m/s) [the pressures are subsequently adjusted for a wind velocity of 101 mi/h (45 m/s)], a mean roof height of 36 ft (11.0 m), and $L/B = 75/50 = 1.5$ (in S.I.: 22.9/15.2 = 1.5).

  At the top of the wall: $p_h = 27.9$ lb/ft² (1.34 kN/m²) (by linear interpolation)

  At the bottom of the wall: $p_o = 25.4$ lb/ft² (1.22 kN/m²) (by linear interpolation)

  These pressures are multiplied by $(101)^2/(110)^2 = 0.84$ [in S.I.: $(45)^2/(49)^2 = 0.84$] to obtain the net pressures for $V = 101$ mi/h (45 m/s).

  At the top of the wall:

  $$p_h = 0.84 \times 27.9 = 23.4 \text{ lb/ft}^2 \text{ (in S.I.}: 0.84 \times 1.34 = 1.13 \text{ kN/m}^2)$$

  At the bottom of the wall:

  $$p_o = 0.84 \times 25.4 = 21.3 \text{ lb/ft}^2 \text{ (in S.I.}: 0.84 \times 1.22 = 1.03 \text{ kN/m}^2)$$

According to note 2 in ASCE/SEI Table 27.5-1, the uniform side wall external pressures are equal to $[0.5(0.54 + 0.64)p_h = 0.59 \times (-23.4) = -13.8$ lb/ft² [in S.I.: $0.59 \times (-1.13) = -0.67$ kN/m²].

These pressures are applied to the projected area of the building walls in the direction of the wind (see ASCE/SEI Figure 27.5-1 and Fig. 3.9).

Step 6—Determine the net wind pressures on the roof   *ASCE/SEI Table 27.5-2*

The wind pressures on the roof are independent of L/B, so the roof pressures for wind in the N-S and E-W directions are the same.

Roof pressures are obtained by reading the values from ASCE/SEI Table 27.5-2 for Exposure C, a wind velocity of 110 mi/h (49 m/s) [the pressures are subsequently adjusted for a wind velocity of 101 mi/h (45 m/s)], and a mean roof height of 36 ft (11.0 m).

The following zones are applicable for a flat roof (roof slope < 9.46 degrees):

- Zone 3: $p_3 = -28.4$ lb/ft² (-1.36 kN/m²) (by linear interpolation) applied normal to the roof area from the windward edge to $0.5h = 18$ ft (5.5 m) from the windward edge.

- Zone 4: $p_4 = -25.4$ lb/ft² (-1.22 kN/m²) (by linear interpolation) applied normal to the roof area from $0.5h = 18$ ft (5.5 m) from the windward edge to $h = 36$ ft (11.0 m) from the windward edge.

- Zone 5: $p_5 = -20.8$ lb/ft² (-1.00 kN/m²) (by linear interpolation) applied normal to the remaining roof area.

These pressures are multiplied by $(101)^2/(110)^2 = 0.84$ [in S.I. : $(45)^2/(49)^2 = 0.84$] to obtain the net pressures for $V = 101$ mi/h (45 m/s):

Zone 3: $p_3 = -0.84 \times 28.4 = -23.9$ lb/ft² (in S.I. : $-0.84 \times 1.36 = -1.14$ kN/m²)

Zone 4: $p_4 = -0.84 \times 25.4 = -21.3$ lb/ft² (in S.I. : $-0.84 \times 1.22 = -1.03$ kN/m²)

Zone 5: $p_5 = -0.84 \times 20.8 = -17.5$ lb/ft² (in S.I. : $-0.84 \times 1.00 = -0.84$ kN/m²)

Step 7—Multiply the net roof pressures by the exposure adjustment factor in ASCE/SEI Table 27.5-2

Net roof pressures must be multiplied by the appropriate exposure adjustment factors for buildings with Exposure B or D. Because the Exposure is C for this building, such an adjustment is not required.

Step 8—Multiply the net wall and roof pressures by K_{zt} and apply the final net pressures on the walls and roof simultaneously

Because $K_{zt} = 1.0$ (see Step 4), the pressures determined in steps 5 and 6 are the final net pressures on the walls and roof, respectively.

Net design wind pressures in the N-S and E-W directions are given in Fig. 3.17.

The building must be designed for the wind load cases defined in ASCE/SEI Figure 27.3-8 (see Fig. 3.8). Distribution of tabulated net wall pressures between windward and leeward wall faces is based on Note 4 in ASCE/SEI Table 27.5-1.

The minimum design wind loading prescribed in ASCE/SEI 27.1.5 must be considered as a load case in addition to the load cases above.

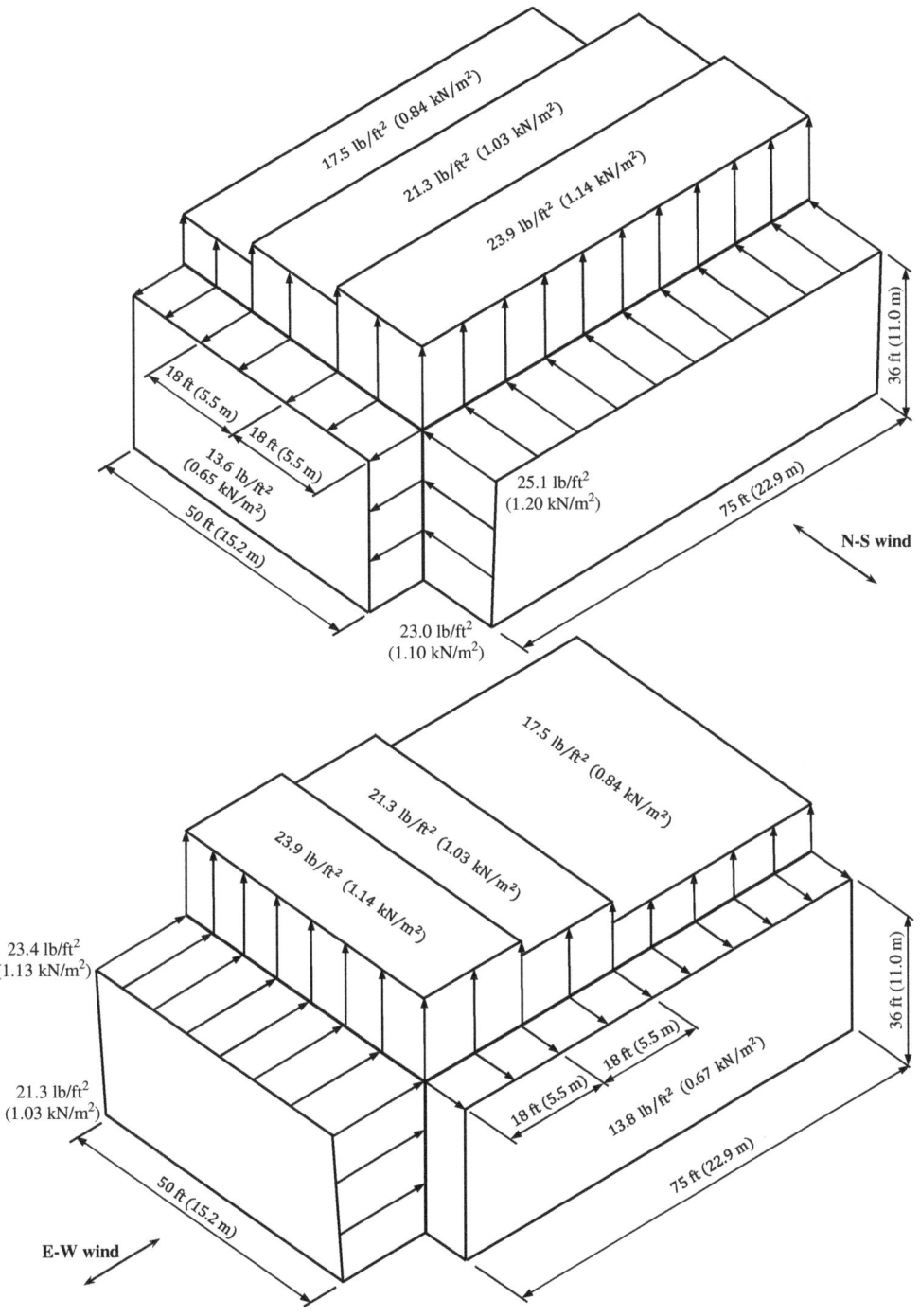

FIGURE 3.17   Net design wind pressures in the N-S and E-W directions for the building in Example 3.2.

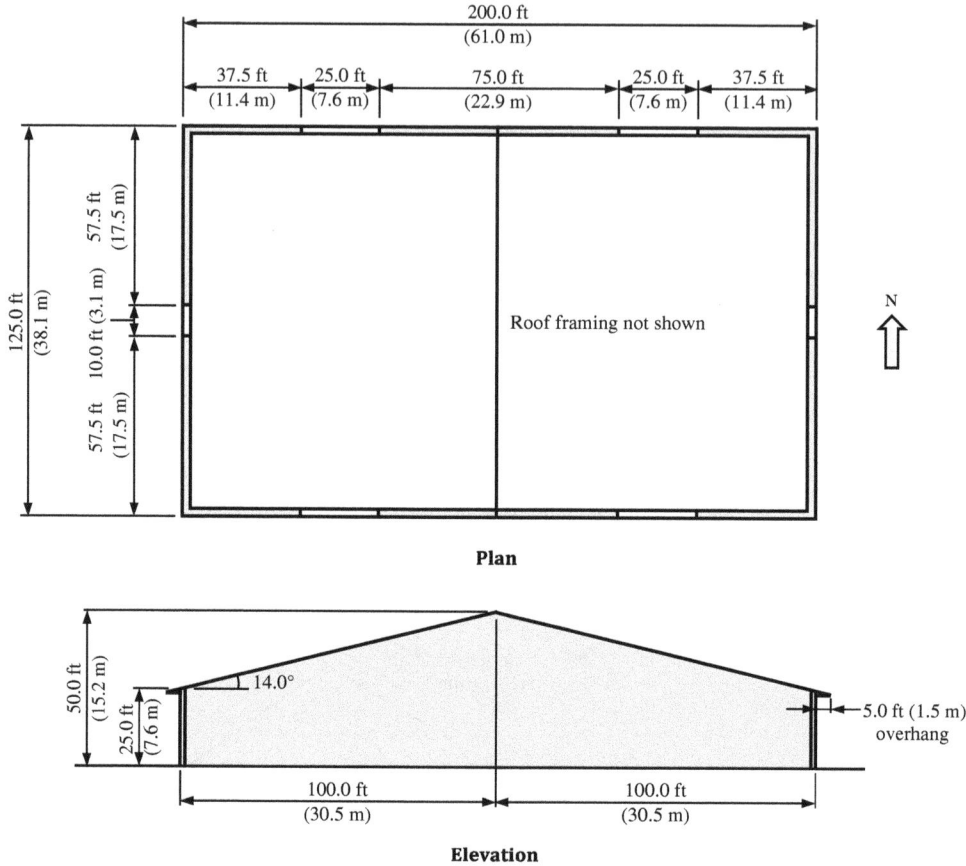

3.4.3   Example 3.3—Wind Pressures on a Retail Building, MWFRS, Chapter 27, Part 1

Determine the wind pressures in both directions on the MWFRS of the one-story retail building in Fig. 3.18 using the requirements in Part 1 of Chapter 27 and the design data in Table 3.7.

Location	Philadelphia, PA
Surface roughness	B
Topography	Not situated on a hill, ridge, or escarpment
Occupancy	Less than 300 people congregate in one area at the same time
Enclosure classification	Enclosed
MWFRS	8-in. (203-mm) thick concrete masonry walls at the perimeter of the building

Tᴀʙʟᴇ **3.7**   Design Data for the Retail Building in Example 3.3

Solution

Check if the building meets all the conditions in ASCE/SEI 27.1.2, so that Part 1 in Chapter 27 can be used to determine the design wind pressures on the MWFRS.

The building is regular-shaped and does not have any unusual geometric irregularities in spatial form. Also, the building does not have any response characteristics that make it subject to across-wind loading or similar effects, and it is not sited at a location where channeling effects or buffeting in the wake of upwind obstructions need to be considered.

Therefore, the conditions in ASCE/SEI 27.1.2 are met and the requirements in Part 1 of Chapter 27 may be used.

The design procedure in Fig. 3.3 is used to determine the design wind pressures, p, in both directions.

Step 1—Determine the velocity pressure, q_z, over the height of the windward wall   Fig. 2.3

The flowchart in Fig. 2.3 is used to determine q_z. For this one-story building, the velocity pressure on the windward wall varies over the height of the wall.

- Step 1a—Determine the surface roughness category   ASCE/SEI 26.7.2

  In Table 3.7, the surface roughness is given as B.

- Step 1b—Determine the exposure category   ASCE/SEI 26.7.3

  It is assumed that surface roughness B applies in all directions and that exposures C and D are not applicable. Therefore, the exposure category is B.

- Step 1c—Determine the terrain exposure constants
       ASCE/SEI Table 26.11-1

  For Exposure B, $\alpha = 7.0$ and $z_g = 1,200$ ft (365.76 m).

- Step 1d—Determine the velocity pressure exposure coefficients, K_z
       ASCE/SEI Table 26.10-1

  Mean roof height $= (50 + 25)/2 = 37.5$ ft (11.4 m)

A summary of the velocity exposure coefficients is given in Table 3.8.

Height above Ground Level, z, ft (m)	K_z
50.0 (15.2)	0.81
37.5 (11.4)	0.75
25.0 (7.6)	0.67
15.0 (4.6)	0.58

TABLE 3.8 Velocity Exposure Coefficients, K_z, for the Building in Example 3.3

For example, at the mean roof height:

$$K_z = K_h = 2.01(z/z_g)^{2/\alpha} = 2.01 \times (37.5/1,200)^{2/7.0} = 0.75$$

In S.I.: $K_z = K_h = 2.01 \times (11.4/365.76)^{2/7.0} = 0.75$

- Step 1e—Determine the topographic factors, K_{zt} ASCE/SEI 26.8

  Because the building is not located on a hill, ridge, or escarpment,

  $$K_{zt} = 1.0.$$

- Step 1f—Determine the wind directionality factor, K_d

                                               ASCE/SEI Table 26.6-1

  For the MWFRS of a building structure, $K_d = 0.85$.

- Step 1g—Determine the ground elevation factor, K_e ASCE/SEI 26.9

  Ground elevation factor can be taken as 1.0 for all elevations.

- Step 1h—Determine the risk category of the building

                                               ASCE/SEI Table 1.5-1

  Due to the nature of its occupancy, this retail building falls under Risk Category II.

- Step 1i—Determine the basic wind speed, V Table 2.1

  For Risk Category II, use IBC Figure 1609.3(1) or ASCE/SEI Figure 26.5-1B. Equivalently, use Ref. 3 or Ref. 4 to obtain $V = 115$ mi/h (51 m/s) for Philadelphia, PA.

- Step 1j—Determine the wind velocity pressures, q_z ASCE/SEI 26.10.2

  A summary of the wind velocity pressures is given in Table 3.9.

  For example, at the mean roof height:

  $$q_z = q_h = 0.00256 K_h K_{zt} K_d K_e V^2$$

  $$= 0.00256 \times 0.75 \times 1.0 \times 0.85 \times 1.0 \times 115^2 = 21.6 \text{ lb/ft}^2$$

  In S.I.:

  $$q_z = q_h = 0.613 K_h K_{zt} K_d K_e V^2$$

  $$= 0.613 \times 0.75 \times 1.0 \times 0.85 \times 1.0 \times 51^2 / 1,000 = 1.02 \text{ kN/m}^2$$

Height above Ground Level, z, ft (m)	q_z, lb/ft² (kN/m²)
50.0 (15.2)	23.3 (1.10)
37.5 (11.4)	21.6 (1.02)
25.0 (7.6)	19.3 (0.91)
15.0 (4.6)	16.7 (0.79)

TABLE 3.9 Wind Velocity Pressures, q_z, for the Building in Example 3.3

Step 2—Determine the velocity pressure, q_h, on the roof and side walls   Fig. 2.3

From Step 1, $q_h = 21.6$ lb/ft^2 (1.02 kN/m^2).

Step 3—Determine the gust-effect factor   *Fig. 2.4*

- Step 3a—Determine the fundamental natural frequency, n_1

  In lieu of obtaining n_1 from a dynamic analysis of the building, check if the approximate lower bound natural frequency, n_a, given in ASCE/SEI 26.11.3 can be used for this building:

  - Building height = 50.0 ft (15.2 m) < 300 ft (91.4 m)
  - Building height = 50.0 ft (15.2 m) < $4L_{eff}$ = 4×125 = 500 ft (152.4 m) in the N-S direction

    = 50.0 ft (15.2 m) < $4L_{eff}$ = 4×200 = 800 ft (243.8 m) in the E-W direction

Because both of these limitations are satisfied, ASCE/SEI Equation (26.11-5) can be used to determine n_a in both directions for the MWFRS consisting of concrete masonry walls.

In the N-S direction:

A_b = base area of the building = 200×125 = 25,000 ft^2 (2,323 m^2)

h = mean roof height = 37.5 ft (11.4 m)

h_i = height of walls = 25 ft (7.6 m)

A_i = horizontal cross-sectional area of wall = (7.625/12)×(125 − 10)

= 73.1 ft^2 (6.8 m^2)

D_i = length of wall = 125 − 10 = 115 ft (35.1 m)

$$C_w = \frac{100}{A_B} \sum_{i=1}^{n} \left(\frac{h}{h_i}\right)^2 \frac{A_i}{\left[1+0.83\left(\frac{h_i}{D_i}\right)^2\right]} = \frac{2\times100}{25,000}\times\left(\frac{37.5}{25.0}\right)^2 \times \frac{73.1}{\left[1+0.83\left(\frac{25.0}{115.0}\right)^2\right]} = 1.27$$

$n_a = 385(C_w)^{0.5}/h = 385\times(1.27)^{0.5}/37.5 = 11.6$ Hz > 1 Hz

In S.I.:

$$C_w = \frac{100}{A_B} \sum_{i=1}^{n} \left(\frac{h}{h_i}\right)^2 \frac{A_i}{\left[1+0.83\left(\frac{h_i}{D_i}\right)^2\right]} = \frac{2\times100}{2,323}\times\left(\frac{11.4}{7.6}\right)^2 \times \frac{6.8}{\left[1+0.83\left(\frac{7.6}{33.5}\right)^2\right]} = 1.27$$

$n_a = 117.3(C_w)^{0.5}/h = 117.3\times(1.27)^{0.5}/11.4 = 11.6$ Hz > 1 Hz

Therefore, the building is rigid in the N-S direction.

Similar calculations in the E-W direction result in n_a > 1 Hz, so the building is rigid in that direction as well.

- Step 3b—Determine the gust-effect factor, G
  For rigid buildings, G is permitted to be taken as 0.85.   ASCE/SEI 26.11.4

Step 4—Determine the external pressure coefficients, C_p   *ASCE/SEI Figure 27.3-1*

- N-S wind
  Windward wall: $C_p = 0.8$
  Leeward wall: $L/B = 125/200 = 0.63$; $C_p = -0.5$
  Side walls: $C_p = -0.7$
  Roof: parallel to the ridge with $h/L = 37.5/125 = 0.30$
  From 0 to 18.8 ft (0 to 5.7 m): $C_p = -0.9, -0.18$
  From 18.8 ft to 37.5 ft (5.7 m to 11.4 m): $C_p = -0.9, -0.18$
  From 37.5 ft to 75 ft (11.4 m to 22.9 m): $C_p = -0.5, -0.18$
  From 75 ft to 125 ft (22.9 m to 38.1 m): $C_p = -0.3, -0.18$
- E-W wind
  Windward wall: $C_p = 0.8$
  Leeward wall: $L/B = 200/125 = 1.6$; $C_p = -0.38$ (by linear interpolation)
  Side walls: $C_p = -0.7$
  Roof: normal to ridge with $\theta = 14.0$ degrees and $h/L = 37.5/200 = 0.19$
  Windward: $C_p = -0.54, -0.04$ (by linear interpolation)
  Leeward: $C_p = -0.46$ (by linear interpolation)
  The roof pressure coefficient $C_p = -0.18$ may become critical where wind loads are combined with roof live loads or snow loads. Determination of wind pressures based on this pressure coefficient should be performed, but such calculations are not shown in this example.

Step 5—Determine the velocity pressure for internal pressure determination, q_i   *Sec. 3.2.1*
  According to ASCE/SEI 27.3.1, $q_i = q_h = 21.6$ lb/ft² (1.02 kN/m²)

Step 6—Determine the enclosure classification   *Table 2.7*
  In the design data, the building is given as enclosed.

Step 7—Determine the internal pressure coefficient, (GC_{pi})   *Table 2.9*
  For an enclosed building, $(GC_{pi}) = +0.18, -0.18$.

Step 8—Determine the design wind pressures, p
- Windward walls
$$p_z = q_z G C_p - q_h (GC_{pi})$$
$$= (q_z \times 0.85 \times 0.8) - [21.6 \times (\pm 0.18)]$$
$$= (0.68 q_z \mp 3.9) \text{ lb/ft}^2$$

In S.I.:

$$p_z = (q_z \times 0.85 \times 0.8) - [1.02 \times (\pm 0.18)] = (0.68 q_z \mp 0.18) \text{ kN/m}^2$$

- Leeward wall, side walls, and roof

$$p_h = q_h GC_p - q_h (GC_{pi})$$

$$= (21.6 \times 0.85 \times C_p) - [21.6 \times (\pm 0.18)]$$

$$= (18.4 C_p \mp 3.9) \text{ lb/ft}^2$$

In S.I.:

$$p_h = (1.02 \times 0.85 \times C_p) - [1.02 \times (\pm 0.18)] = (0.87 C_p \mp 0.18) \text{ kN/m}^2$$

Design wind pressures for wind in the N-S and E-W directions are given in Tables 3.10 and 3.11, respectively.

Building Surface	Height above Ground Level, z, ft (m)	q, lb/ft² (kN/m²)	External Pressure qGC_p, lb/ft² (kN/m²)	Internal Pressure $q_h(GC_{pi})$, lb/ft² (kN/m²)	Net Pressure, p, lb/ft² (kN/m²)	
					(+GC_{pi})	(−GC_{pi})
Windward wall	50.0 (15.2)	23.3 (1.10)	15.8 (0.75)	±3.9 (±0.18)	11.9 (0.57)	19.7 (0.93)
	37.5 (11.4)	21.6 (1.02)	14.7 (0.69)	±3.9 (±0.18)	10.8 (0.51)	18.6 (0.87)
	25.0 (7.6)	19.3 (0.91)	13.1 (0.62)	±3.9 (±0.18)	9.2 (0.44)	17.0 (0.80)
	15.0 (4.6)	16.7 (0.79)	11.4 (0.54)	±3.9 (±0.18)	7.5 (0.36)	15.3 (0.72)
Leeward wall	All	21.6 (1.02)	−9.2 (−0.44)	±3.9 (±0.18)	−13.1 (−0.62)	−5.3 (−0.26)
Side walls	All	21.6 (1.02)	−12.9 (−0.61)	±3.9 (±0.18)	−16.8 (−0.79)	−9.0 (−0.43)
Roof	___(1)	21.6 (1.02)	−16.6 (−0.78)	±3.9 (±0.18)	−20.5 (−0.96)	−12.7 (−0.60)
	___(2)	21.6 (1.02)	−16.6 (−0.78)	±3.9 (±0.18)	−20.5 (−0.96)	−12.7 (−0.60)
	___(3)	21.6 (1.02)	−9.2 (−0.44)	±3.9 (±0.18)	−13.1 (−0.62)	−5.3 (−0.26)
	___(4)	21.6 (1.02)	−5.5 (−0.26)	±3.9 (±0.18)	−9.4 (−0.44)	−1.6 (−0.08)

(1) From windward edge of roof to 18.8 ft (5.7 m).
(2) From 18.8 ft (5.7 m) to 37.5 ft (11.4 m).
(3) From 37.5 ft (11.4 m) to 75 ft (22.9 m).
(4) From 75 ft (22.9 m) to 125 ft (38.1 m).

TABLE 3.10 Design Wind Pressures, p, for Wind in the N-S Direction for the Building in Example 3.3

Building Surface	Height above Ground Level, z, ft (m)	q, lb/ft² (kN/m²)	External Pressure qGC_p, lb/ft² (kN/m²)	Internal Pressure q_h(GC_pi), lb/ft² (kN/m²)	Net Pressure, p, lb/ft² (kN/m²) +(GC_pi)	−(GC_pi)
Windward wall	25.0 (7.6)	19.3 (0.91)	13.1 (0.62)	±3.9 (±0.18)	9.2 (0.44)	17.0 (0.80)
	15.0 (4.6)	16.7 (0.79)	11.4 (0.54)	±3.9 (±0.18)	7.5 (0.36)	15.3 (0.72)
Leeward wall	All	21.6 (1.02)	−7.0 (−0.33)	±3.9 (±0.18)	−10.9 (−0.51)	−3.1 (−0.15)
Side walls	All	21.6 (1.02)	−12.9 (−0.61)	±3.9 (±0.18)	−16.8 (−0.79)	−9.0 (−0.43)
Roof	___(1)	21.6 (1.02)	−9.9 (−0.47)	±3.9 (±0.18)	−13.8 (−0.65)	−6.0 (−0.29)
	___(2)	21.6 (1.02)	−0.7 (−0.04)	±3.9 (±0.18)	−4.6 (−0.22)	3.2 (0.14)
	___(3)	21.6 (1.02)	−8.5 (−0.40)	±3.9 (±0.18)	−12.4 (−0.58)	−4.6 (−0.22)

(1) Windward roof with external roof pressure coefficient $C_p = -0.54$.
(2) Windward roof with external roof pressure coefficient $C_p = -0.04$.
(3) Leeward roof.

TABLE 3.11   Design Wind Pressures, p, for Wind in the E-W Direction for the Building in Example 3.3

The positive external pressure on the bottom surface of the windward roof over-hang is determined in accordance with ASCE/SEI 27.3.3 (see Fig. 3.5). The velocity pressure, q_z, at the top of the wall [$z = 25$ ft (7.6 m)] is equal to 19.3 lb/ft² (0.91 kN/m²) (see Table 3.11).

Therefore,

$$p = 0.8q_z G = 0.8 \times 19.3 \times 0.85 = 13.1 \text{ lb/ft}^2$$

In S.I.:

$$p = 0.8q_z G = 0.8 \times 0.91 \times 0.85 = 0.62 \text{ kN/m}^2$$

Net design wind pressures in the N-S and E-W directions for positive and negative internal pressures are given in Figs. 3.19 and 3.20, respectively. For wind in the E-W direction, the wind pressure depicted in Fig. 3.20 on the windward roof is for the case where the external roof pressure coefficient $C_p = -0.54$; a similar pressure diagram can be obtained using $C_p = -0.04$.

The horizontal wind pressures on the walls for both directions are shown in Fig. 3.21. It is evident from Figs. 3.19, 3.20, and 3.21 that the internal pressures cancel out when determining the horizontal wind pressures on the walls.

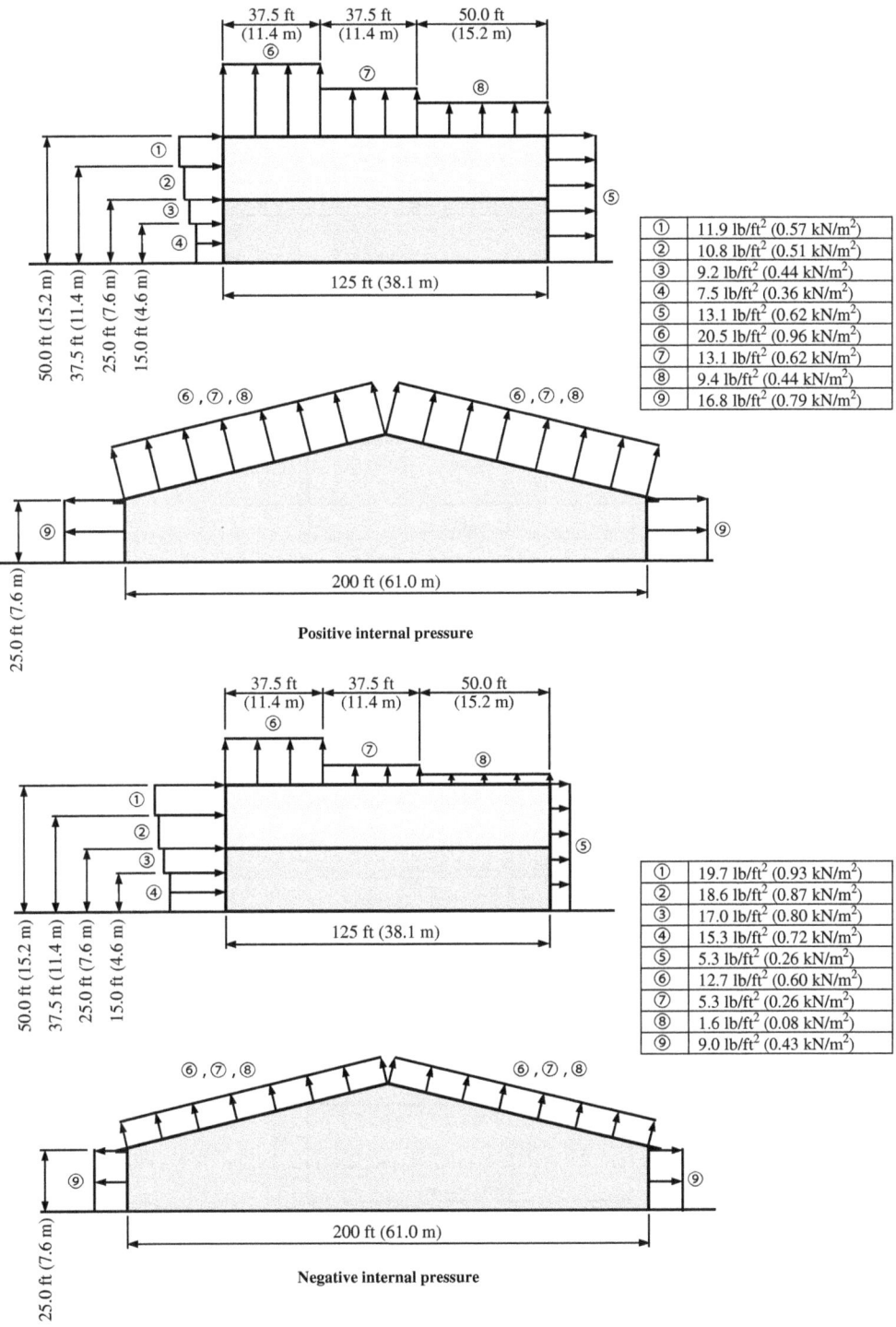

FIGURE 3.19 Net design wind pressures in the N-S direction for the building in Example 3.3.

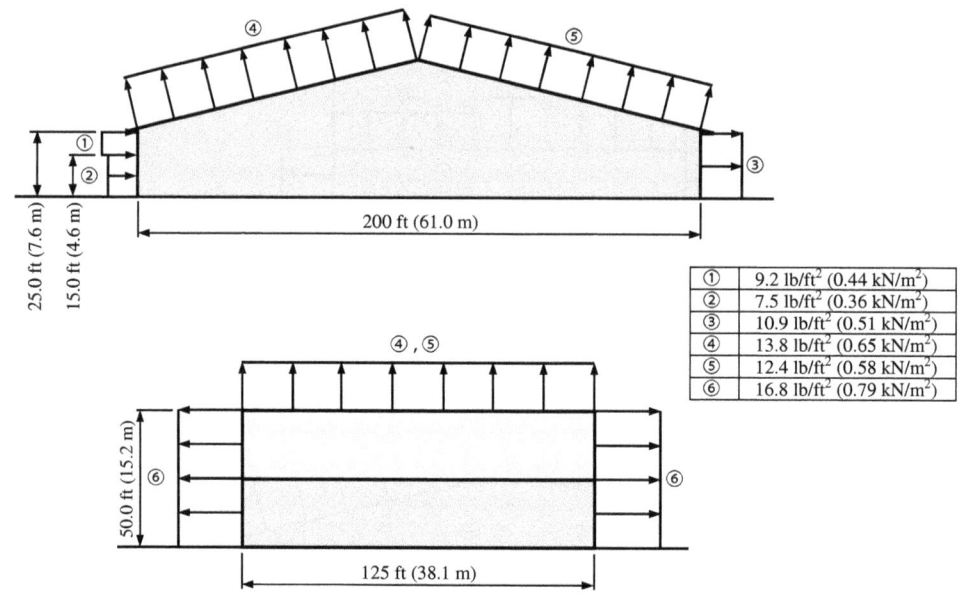

①	9.2 lb/ft² (0.44 kN/m²)
②	7.5 lb/ft² (0.36 kN/m²)
③	10.9 lb/ft² (0.51 kN/m²)
④	13.8 lb/ft² (0.65 kN/m²)
⑤	12.4 lb/ft² (0.58 kN/m²)
⑥	16.8 lb/ft² (0.79 kN/m²)

Positive internal pressure

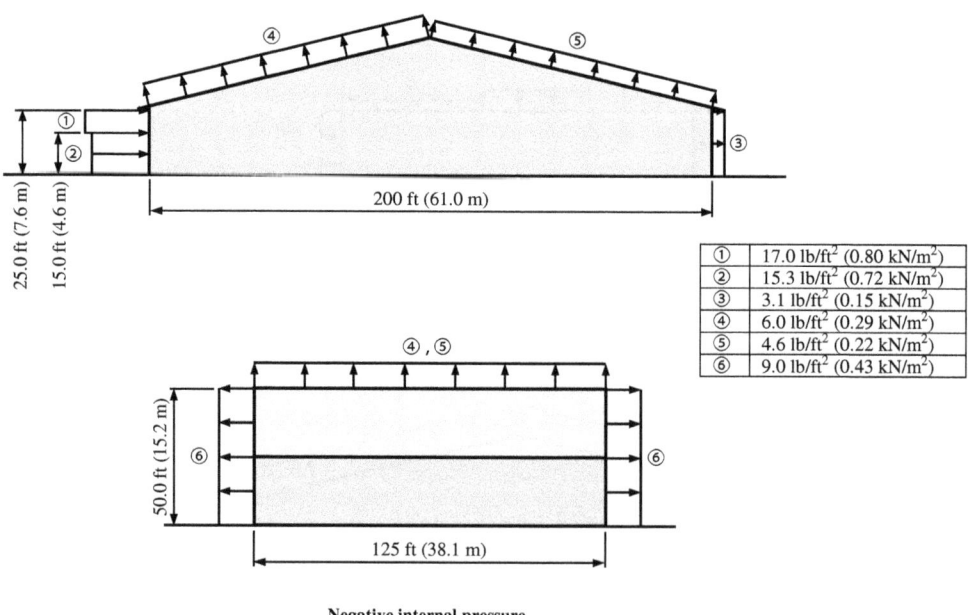

①	17.0 lb/ft² (0.80 kN/m²)
②	15.3 lb/ft² (0.72 kN/m²)
③	3.1 lb/ft² (0.15 kN/m²)
④	6.0 lb/ft² (0.29 kN/m²)
⑤	4.6 lb/ft² (0.22 kN/m²)
⑥	9.0 lb/ft² (0.43 kN/m²)

Negative internal pressure

Figure 3.20   Net design wind pressures in the E-W direction for the building in Example 3.3.

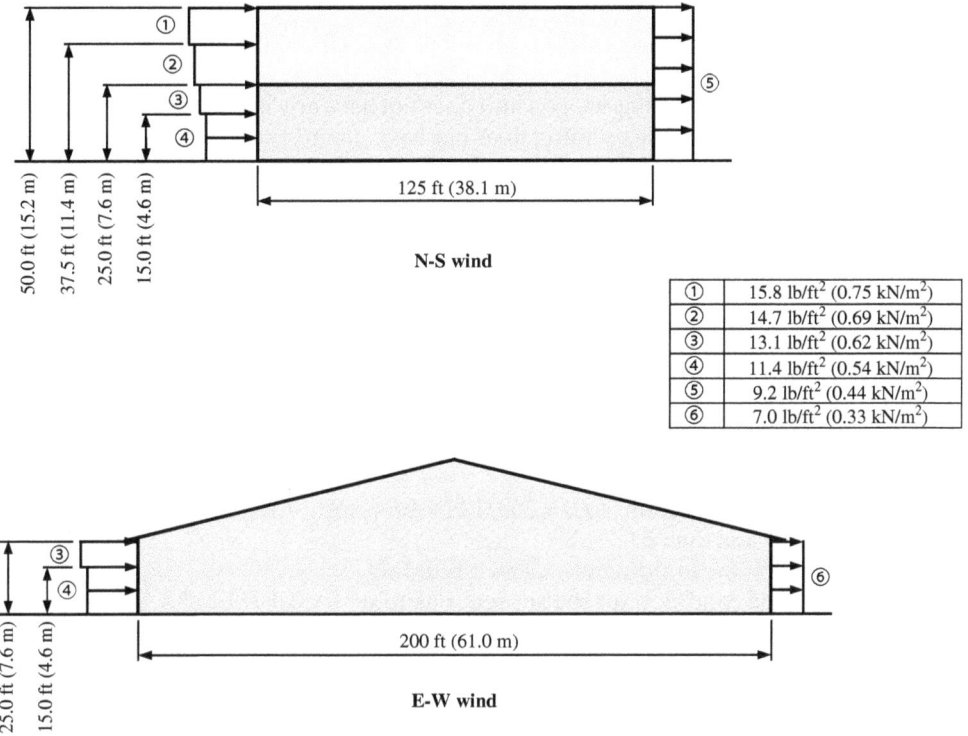

①	15.8 lb/ft^2 (0.75 kN/m^2)
②	14.7 lb/ft^2 (0.69 kN/m^2)
③	13.1 lb/ft^2 (0.62 kN/m^2)
④	11.4 lb/ft^2 (0.54 kN/m^2)
⑤	9.2 lb/ft^2 (0.44 kN/m^2)
⑥	7.0 lb/ft^2 (0.33 kN/m^2)

FIGURE 3.21 Horizontal wind pressures on the walls for the building in Example 3.3.

The building must be designed for the wind load cases defined in ASCE/SEI Figure 27.3-8 (see Fig. 3.8).

In case 1, the full design wind pressures act on the projected area perpendicular to each principal axis of the building at each level above ground. These pressures act separately along each principal axis. The windward and leeward pressures in Fig. 3.21 fall under case 1.

According to the exception in ASCE/SEI 27.3.5, buildings that meet the requirements of ASCE/SEI Appendix D.1 need only be designed for cases 1 and 3. It can be shown that this building is classified as a torsionally regular building under wind loads, which satisfies the requirements in ASCE/SEI D.4. Therefore, only cases 1 and 3 must be considered.

In case 3, 75 percent of the windward and leeward wall pressures in Fig. 3.21 act simultaneously on the building at each level above ground.

The minimum design wind loading prescribed in ASCE/SEI 27.1.5 must be considered as a load case in addition to the load cases above.

3.4.4 Example 3.4—Wind Pressures on a Retail Building, MWFRS, Chapter 27, Part 2

Determine the wind pressures in both directions on the MWFRS of the one-story retail building in Fig. 3.18 using the requirements in Part 2 of Chapter 27 and the design data in Table 3.7. Assume the roof diaphragm is flexible, which transfers the lateral load effects to the concrete masonry walls.

Solution

Check if the building meets all the conditions in ASCE/SEI 27.1.2, 27.4.2, and 27.4.5 so that Part 2 in Chapter 27 can be used to determine the design wind pressures on the MWFRS.

The building is regular-shaped and does not have any unusual geometric irregularities in spatial form. Also, the building does not have any response characteristics that make it subject to across-wind loading or similar effects, and it is not sited at a location where channeling effects or buffeting in the wake of upwind obstructions need to be considered.

Check the conditions for a Class 1 building (ASCE/SEI 27.4.2):

1. The building is enclosed and is a simple diaphragm building as defined in ASCE/SEI 26.2 where the windward and leeward wind loads are transmitted through the continuous flexible diaphragm to the concrete masonry walls (MWFRS).

2. Mean roof height $h = 37.5$ ft (11.4 m) < 60 ft (18.3 m).

3. In the N-S direction, $L/B = 125/200 = 0.63$ (in S.I.: $38.1/61.0 = 0.63$), which is greater than 0.2 and less than 5.0.

In the E-W direction, $L/B = 200/125 = 1.6$ (in S.I.: $61.0/38.1 = 1.6$), which is greater than 0.2 and less than 5.0.

Therefore, the building is a Class 1 building.

Check the condition for diaphragm flexibility (ASCE/SEI 27.4.5):

As noted in the design data, the roof diaphragm is classified as flexible, which is permitted in accordance with ASCE/SEI 27.4.5.

Therefore, the conditions in ASCE/SEI 27.1.2, 27.4.2, and 27.4.5 are met and the requirements in Part 2 of Chapter 27 may be used.

The design procedure in Fig. 3.10 is used to determine the design wind pressures, p, in both directions.

Step 1—Determine the risk category of the building   *ASCE/SEI Table 1.5-1*

Due to the nature of its occupancy, this retail building falls under Risk Category II.

Step 2—Determine the basic wind speed, V   *Table 2.1*

For Risk Category II, use IBC Figure 1609.3(1) or ASCE/SEI Figure 26.5-1B.

Equivalently, use Ref. 3 or Ref. 4 to obtain $V = 115$ mi/h (51 m/s) for Philadelphia, PA.

Step 3—Determine the exposure category   *ASCE/SEI 26.7.3*

It is assumed that surface roughness B applies in all directions and that exposures C and D are not applicable. Therefore, the exposure category is B.

Step 4—Determine the topographic factor, K_{zt}, calculated at 0.33h   *ASCE/SEI 26.8*

Because the building is not located on a hill, ridge, or escarpment, $K_{zt} = 1.0$.

Step 5—Determine the net wind pressures p_h and p_o on the walls   *ASCE/SEI Table 27.5-1*

- N-S wind

   The along-wind net wall pressures are obtained by reading the values from ASCE/SEI Table 27.5-1 for Exposure B, a wind velocity of 115 mi/h (51 m/s), a mean roof height of 37.5 ft (11.4 m), and $L/B = 125/200 = 0.63$ (in S.I.: $38.1/61.0 = 0.63$).

At the top of the wall: $p_h = 23.0$ lb/ft^2 (1.10 kN/m^2) (by linear interpolation)

At the bottom of the wall: $p_o = 20.3$ lb/ft^2 (0.97 kN/m^2) (by linear interpolation)

According to note 2 in ASCE/SEI Table 27.5-1, the uniform side wall external pressures are equal to $0.54p_h = 0.54 \times (-23.0) = -12.4$ lb/ft^2 [in S.I.: $0.54 \times (-1.10) = -0.59$ kN/m^2].

- E-W wind

  The along-wind net wall pressures are obtained by reading the values from ASCE/SEI Table 27.5-1 for Exposure B, a wind velocity of 115 mi/h (51 m/s), a mean roof height of 37.5 ft (11.4 m), and $L/B = 200/125 = 1.6$ (in S.I.: $61.0/38.1 = 1.6$).

  At the top of the wall: $p_h = 21.1$ lb/ft^2 (1.01 kN/m^2) (by linear interpolation)

  At the bottom of the wall: $p_o = 18.5$ lb/ft^2 (0.89 kN/m^2) (by linear interpolation)

  According to note 2 in ASCE/SEI Table 27.5-1, the uniform side wall external pressures are equal to $0.6p_h = 0.6 \times (-21.1) = -12.7$ lb/ft^2 (by linear interpolation).

  In S.I.: $0.6 \times (-1.01) = -0.61$ kN/m^2 (by linear interpolation).

  These pressures are applied to the projected area of the building walls in the direction of the wind (see ASCE/SEI Figure 27.5-1 and Fig. 3.9).

Step 6—Determine the net wind pressures on the roof *ASCE/SEI Table 27.5-2*

Roof pressures are obtained by reading the values from ASCE/SEI Table 27.5-2 for Exposure C, a wind velocity of 115 mi/h (51 m/s) (the pressures are subsequently adjusted for Exposure B), and a mean roof height of 37.5 ft (11.4 m). The following pressures are for load case 1 (according to note 2 in ASCE/SEI Table 27.5-2, load case 2 is required to investigate maximum overturning on the building).

- N-S wind (parallel to ridge)

  The following zones are applicable for a gable roof with a 14.0-degree slope:

  - Zone 3: $p_3 = -31.4$ lb/ft^2 (−1.50 kN/m^2) (by linear interpolation) applied normal to the roof area from the windward edge to $0.5h = 18.8$ ft (5.7 m) from the windward edge.

  - Zone 4: $p_4 = -28.0$ lb/ft^2 (−1.34 kN/m^2) (by linear interpolation) applied normal to the roof area from $0.5h = 18.8$ ft (5.7 m) from the windward edge to $h = 37.5$ ft (11.4 m) from the windward edge.

  - Zone 5: $p_5 = -23.0$ lb/ft^2 (−1.10 kN/m^2) (by linear interpolation) applied normal to the remaining roof area.

- E-W wind (normal to ridge)

  The following zones are applicable for a gable roof with a 14.0-degree slope:

  - Zone 1: $p_1 = -30.8$ lb/ft^2 (−1.48 kN/m^2) (by linear interpolation) applied normal to the windward roof area.

  - Zone 2: $p_2 = -22.2$ lb/ft^2 (−1.06 kN/m^2) (by linear interpolation) applied normal to the leeward roof area.

Step 7—Multiply the net roof pressures by the exposure adjustment factor in ASCE/SEI Table 27.5-2

Net roof pressures must be multiplied by the appropriate exposure adjustment factors for buildings with Exposure B or D. Because the Exposure is B for this building, such an adjustment is required.

From ASCE/SEI Table 27.5-2, the adjustment factor is equal to 0.725 for Exposure B and a mean roof height of 37.5 ft (11.4 m) (by linear interpolation). Therefore,

$$p_1 = 0.725 \times (-30.8) = -22.3 \text{ lb/ft}^2 \; (-1.07 \text{ kN/m}^2)$$

$$p_2 = 0.725 \times (-22.2) = -16.1 \text{ lb/ft}^2 \; (-0.77 \text{ kN/m}^2)$$

$$p_3 = 0.725 \times (-31.4) = -22.8 \text{ lb/ft}^2 \; (-1.09 \text{ kN/m}^2)$$

$$p_4 = 0.725 \times (-28.0) = -20.3 \text{ lb/ft}^2 \; (-0.97 \text{ kN/m}^2)$$

$$p_5 = 0.725 \times (-23.0) = -16.7 \text{ lb/ft}^2 \; (-0.80 \text{ kN/m}^2)$$

Step 8—Multiply the net wall and roof pressures by K_{zt} and apply the final net pressures on the walls and roof simultaneously

Because $K_{zt} = 1.0$ (see step 4), the pressures determined in steps 5 and 7 are the final net pressures on the walls and roof.

Net design wind pressures in the N-S and E-W directions are given in Fig. 3.22.

The building must be designed for the wind load cases defined in ASCE/SEI Figure 27.3-8 (see Fig. 3.8). Distribution of tabulated net wall pressures between windward and leeward wall faces is based on note 4 in ASCE/SEI Table 27.5-1.

The minimum design wind loading prescribed in ASCE/SEI 27.1.5 must be considered as a load case in addition to the load cases above.

3.4.5 Example 3.5—Wind Pressures on an Essential Facility, MWFRS, Chapter 27, Part 1

Determine the wind pressures in both directions on the MWFRS of the essential facility in Fig. 3.23 using the requirements in Part 1 of Chapter 27 and the design data in Table 3.12.

Solution
Check if the building meets all the conditions in ASCE/SEI 27.1.2 so that Part 1 in Chapter 27 can be used to determine the design wind pressures on the MWFRS.

The building is regular-shaped and does not have any unusual geometric irregularities in spatial form. Also, the building does not have any response characteristics that make it subject to across-wind loading or similar effects, and it is not sited at a location where channeling effects or buffeting in the wake of upwind obstructions need to be considered.

Therefore, the conditions in ASCE/SEI 27.1.2 are met and the requirements in Part 1 of Chapter 27 may be used.

The design procedure in Fig. 3.3 is used to determine the design wind pressures, *p*, in both directions.

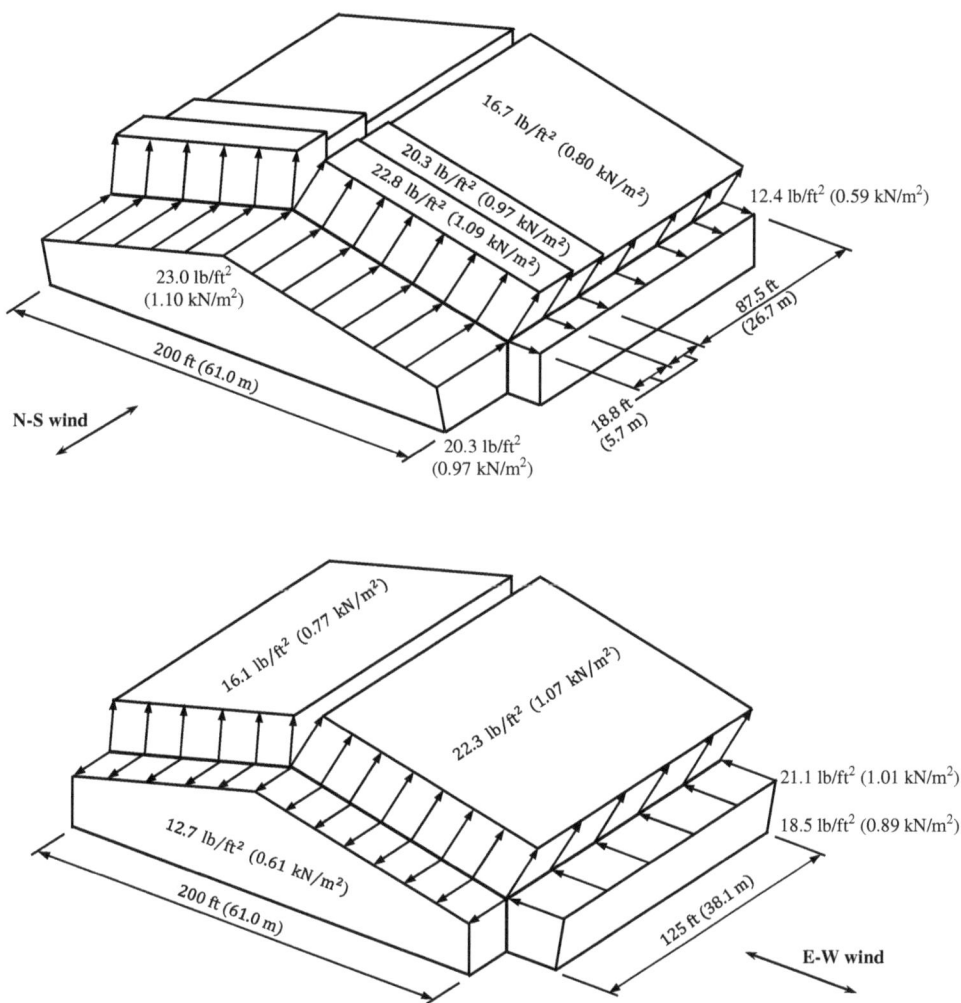

FIGURE 3.22 Net design wind pressures in the N-S and E-W directions for the building in Example 3.4.

Step 1—Determine the velocity pressure, q_z, over the height of the windward wall   *Fig. 2.3*

The flowchart in Fig. 2.3 is used to determine q_z.

- Step 1a—Determine the surface roughness category   ASCE/SEI 26.7.2

  In Table 3.12, the surface roughness is given as C.

- Step 1b—Determine the exposure category   ASCE/SEI 26.7.3

  It is assumed that surface roughness C applies in all directions and that exposures B and D are not applicable. Therefore, the exposure category is C.

- Step 1c—Determine the terrain exposure constants   ASCE/SEI Table 26.11-1

  For Exposure C, $\alpha = 9.5$ and $z_g = 900$ ft (274.32 m).

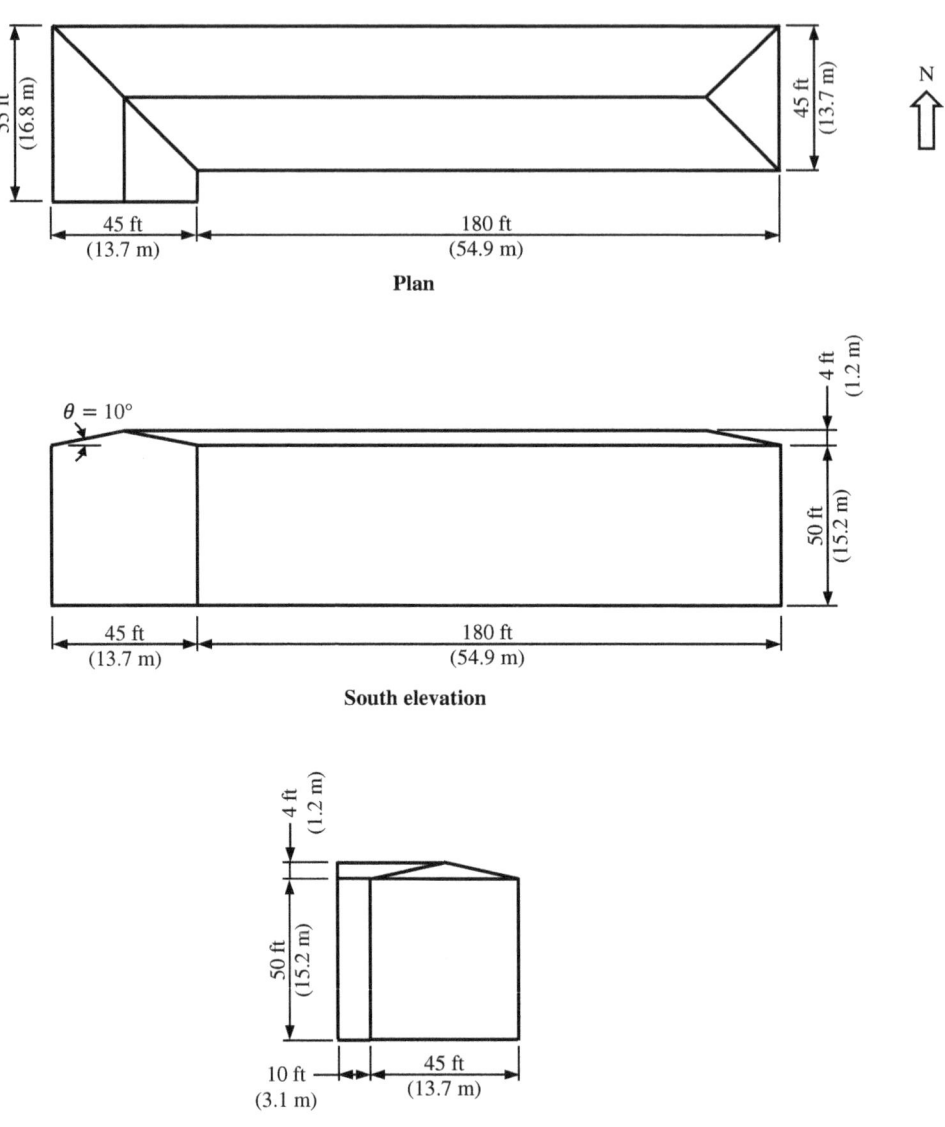

Figure 3.23 Plan and elevations of the essential facility in Example 3.5.

Location	Charleston, SC
Surface roughness	C
Topography	Not situated on a hill, ridge, or escarpment
Occupancy	Essential facility that must remain operational at all times
Enclosure classification	Enclosed
Fundamental natural frequency of building structure	>1 Hz in both principal directions

Table 3.12 Design Data for the Essential Facility in Example 3.5

- Step 1d—Determine the velocity pressure exposure coefficient, K_z
  ASCE/SEI Table 26.10-1

$$\text{Mean roof height, } h = \frac{50+54}{2} = 52 \text{ ft (15.9 m)}$$

Values of K_z are given in Table 3.13 for Exposure C.

For example, at $z = 52$ ft (15.9 m):

$$K_z = 2.01(z/z_g)^{2/\alpha} = 2.01 \times (52/900)^{2/9.5} = 1.10$$

In S.I.:

$$K_z = 2.01 \times (15.9/274.32)^{2/9.5} = 1.10$$

- Step 1e—Determine the topographic factors, K_{zt} ASCE/SEI 26.8
  Because the building is not located on a hill, ridge, or escarpment, $K_{zt} = 1.0$.
- Step 1f—Determine the wind directionality factor, K_d ASCE/SEI Table 26.6-1
  For the MWFRS of a building structure, $K_d = 0.85$.
- Step 1g—Determine the ground elevation factor, K_e ASCE/SEI 26.9
  Ground elevation factor can be taken as 1.0 for all elevations.
- Step 1h—Determine the risk category of the building   ASCE/SEI Table 1.5-1
  Due to the nature of its occupancy, this essential facility falls under Risk Category IV.
- Step 1i—Determine the basic wind speed, V Table 2.1
  For Risk Category IV, use IBC Figure 1609.3(3) or ASCE/SEI Figure 26.5-1D.
  Equivalently, use Ref. 3 or Ref. 4 to obtain $V = 164$ mi/h (73 m/s) for Charleston, SC.
- Step 1j—Determine the wind velocity pressure, q_z ASCE/SEI 26.10.2

$$q_z = 0.00256 K_z K_{zt} K_d K_e V^2$$

$$= 0.00256 \times K_z \times 1.0 \times 0.85 \times 1.0 \times 164^2 = 58.5 K_z \text{ lb/ft}^2$$

Height above Ground Level, z, ft (m)	K_z
52 (15.9)	1.10
40 (12.2)	1.04
30 (9.1)	0.98
20 (6.1)	0.90
10 (3.1)	0.85

TABLE 3.13 Velocity Pressure Exposure Coefficient, K_z, for the Essential Facility in Example 3.5

In S.I.:

$$q_z = 0.613 K_z K_{zt} K_d K_e V^2$$
$$= 0.613 \times K_z \times 1.0 \times 0.85 \times 1.0 \times 73^2 / 1,000 = 2.78 K_z \text{ kN/m}^2$$

The velocity pressures over the height of the windward wall are given in Table 3.14.

Step 2—Determine the velocity pressure, q_h, on the roof and side walls   Fig. 2.3

From step 1, $q_h = 64.4$ lb/ft² (3.06 kN/m²).

Step 3—Determine the gust-effect factor   Fig. 2.4

- Step 3a—Determine the fundamental natural frequency, n_1

The fundamental natural frequency, n_1, of the building structure is given in the design data as greater than 1 Hz in both principal directions.

Therefore, the building is rigid in both directions.

- Step 3b—Determine the gust-effect factor, G

For rigid buildings, G is permitted to be taken as 0.85.   ASCE/SEI 26.11.4

- Step 4—Determine the external pressure coefficients, C_p

ACE/SEI Figure 27.3-1

Because the building is not symmetric, all four wind directions normal to the walls must be considered.

Identifications marks for each surface of the building are given in Fig. 3.24.

External pressure coefficients, C_p, for wind in all four directions are given in Tables 3.15 through 3.18.

Step 5—Determine the velocity pressure for internal pressure determination, q_i

Sec. 3.2.1

According to ASCE/SEI 27.3.1, $q_i = q_h = 64.4$ lb/ft² (3.06 kN/m²)

Height above Ground Level, z, ft (m)	K_z	q_z, lb/ft² (kN/m²)
52 (15.9)	1.10	64.4 (3.06)
40 (12.2)	1.04	60.8 (2.89)
30 (9.1)	0.98	57.3 (2.72)
20 (6.1)	0.90	52.7 (2.50)
10 (3.1)	0.85	49.7 (2.36)

TABLE 3.14 Velocity Pressure, q_z, for the Essential Facility in Example 3.5

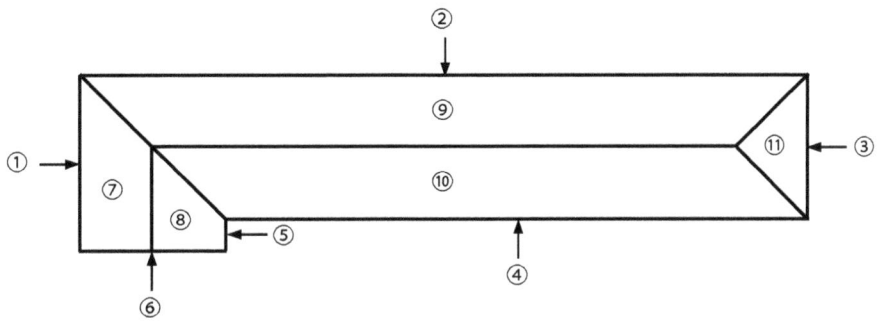

Figure 3.24 Identification marks for the building surfaces of the essential facility in Example 3.5.

Step 6—Determine the enclosure classification   Table 2.7

In the design data, the building is given as enclosed. Also, because the building is located in a hurricane-prone region, glazed openings must be protected against windborne debris in accordance with ASCE/SEI 26.12.3.

Step 7—Determine the internal pressure coefficient, (GC_{pi})   Table 2.9

For an enclosed building, $(GC_{pi}) = +0.18, -0.18$.

Surface(s)	Type		C_p
1, 3, 5	Side wall		−0.70
2	Windward wall		0.80
4, 6	Leeward wall[1]		−0.50
7, 8	Roof parallel to ridge[2]	Windward edge to 26 ft (7.9 m)	−0.90, −0.18
		26 ft (7.9 m) to 52 ft (15.9 m)	−0.90, −0.18
		52 ft (15.9 m) 55 ft (16.8 m)	−0.50, −0.18
9	Windward roof[3]		−1.26, −0.18
10	Leeward roof[3]		−0.68
11	Roof parallel to ridge[2]	Windward edge to 26 ft (7.9 m)	−0.90, −0.18
		26 ft (7.9 m) to 45 ft (13.7 m)	−0.90, −0.18

[1]Obtained using $L/B = 55/225 = 0.24$ (in S.I.: $L/B = 16.8/68.6 = 0.24$).
[2]The smaller uplift pressures on the roof due to $C_p = -0.18$ may govern the design when combined with roof live load or snow loads. This pressure is not considered in this example, but in general should be considered.
[3]Obtained by linear interpolation using $\theta = 10$ deg and $h/L = 52/55 = 0.95$ (in S.I.: $h/L = 15.9/16.8 = 0.95$).

Table 3.15 External Pressure Coefficients, C_p, for Wind from North to South

Surface(s)	Type		C_p
1, 3, 5	Side wall		−0.70
2	Leeward wall[1]		−0.50
4, 6	Windward wall		0.80
7, 8	Roof parallel to ridge[2]	Windward edge to 26 ft (7.9 m)	−0.90, −0.18
		26 ft (7.9 m) to 52 ft (15.9 m)	−0.90, −0.18
		52 ft (15.9 m) 55 ft (16.8 m)	−0.50, −0.18
9	Leeward roof[3]		−0.68
10	Windward roof[3]		−1.26, −0.18
11	Roof parallel to ridge[2]	Windward edge to 26 ft (7.9 m)	−0.90, −0.18
		26 ft (7.9 m) to 45 ft (13.7 m)	−0.90, −0.18

[1]Obtained using $L/B = 55/225 = 0.24$ (in S.I.: $L/B = 16.8/68.6 = 0.24$).
[2]The smaller uplift pressures on the roof due to $C_p = -0.18$ may govern the design when combined with roof live load or snow loads. This pressure is not considered in this example, but in general should be considered.
[3]Obtained by linear interpolation using $\theta = 10$ degree and $h/L = 52/55 = 0.95$ (in S.I.: $h/L = 15.9/16.8 = 0.95$).

TABLE 3.16 External Pressure Coefficients, C_p, for Wind from South to North

Step 8—Determine the design wind pressures, p

- Windward walls for wind in both directions

$$p_z = q_z GC_p - q_h(GC_{pi})$$
$$= (q_z \times 0.85 \times 0.8) - [64.4 \times (\pm 0.18)]$$
$$= (0.68q_z \mp 11.6) \text{ lb/ft}^2$$

Surface(s)	Type		C_p
1	Leeward wall[1]		−0.20
2, 4, 6	Side wall		−0.70
3, 5	Windward wall		0.80
7	Leeward roof[2]		−0.30
8, 11	Windward roof[2]		−0.70, −0.18
9, 10	Roof parallel to ridge[3]	Windward edge to 26 ft (7.9 m)	−0.90, −0.18
		26 ft (7.9 m) to 52 ft (15.9 m)	−0.90, −0.18
		52 ft (15.9 m) 104 ft (31.7 m)	−0.50, −0.18
		104 ft (31.7 m) to 225 ft (68.6 m)	−0.30, −0.18

[1]Obtained using $L/B = 225/55 = 4.1$ (in S.I.: $L/B = 68.6/16.8 = 4.1$).
[2]Obtained using $\theta = 10$ degree and $h/L = 52/225 = 0.23$ (in S.I.: $h/L = 15.9/68.6 = 0.23$).
[3]The smaller uplift pressures on the roof due to $C_p = -0.18$ may govern the design when combined with roof live load or snow loads. This pressure is not considered in this example, but in general should be considered.

TABLE 3.17 External Pressure Coefficients, C_p, for Wind from East to West

Surface(s)	Type		C_p
1	Windward wall		0.80
2, 4, 6	Side wall		−0.70
3, 5	Leeward wall[(1)]		−0.20
7	Windward roof[(2)]		−0.70, −0.18
8, 11	Leeward roof[(2)]		−0.30
9, 10	Roof parallel to ridge[(3)]	Windward edge to 26 ft (7.9 m)	−0.90, −0.18
		26 ft (7.9 m) to 52 ft (15.9 m)	−0.90, −0.18
		52 ft (15.9 m) 104 ft (31.7 m)	−0.50, −0.18
		104 ft (31.7 m) to 225 ft (68.6 m)	−0.30, −0.18

[(1)]Obtained using $L/B = 225/55 = 4.1$ (in S.I.: $L/B = 68.6/16.8 = 4.1$).
[(2)]Obtained using $\theta = 10$ degree and $h/L = 52/225 = 0.23$ (in S.I.: $h/L = 15.9/68.6 = 0.23$).
[(3)]The smaller uplift pressures on the roof due to $C_p = -0.18$ may govern the design when combined with roof live load or snow loads. This pressure is not considered in this example, but in general should be considered.

TABLE 3.18 External Pressure Coefficients, C_p, for Wind from West to East

In S.I.:

$$p_z = (q_z \times 0.85 \times 0.8) - [3.06 \times (\pm 0.18)] = (0.68 q_z \mp 0.55) \text{ kN/m}^2$$

- Leeward wall, side walls, and roof

$$p_h = q_h GC_p - q_h(GC_{pi})$$
$$= (64.4 \times 0.85 \times C_p) - [64.4 \times (\pm 0.18)]$$
$$= (54.7 C_p \mp 11.6) \text{ lb/ft}^2$$

In S.I.:

$$p_h = (3.06 \times 0.85 \times C_p) - [3.06 \times (\pm 0.18)] = (2.60 C_p \mp 0.55) \text{ kN/m}^2$$

Design wind pressures for wind in all four directions are given in Tables 3.19 through 3.22.

The building must be designed for the wind load cases defined in ASCE/SEI Figure 27.3-8 (see Fig. 3.8). Because the building is not symmetrical, all four wind directions must be considered when combining loads in accordance with ASCE/SEI Figure 27.3-8.

In case 1, the full design wind pressures act on the projected area perpendicular to each principal axis of the building at each level above ground. These pressures act separately along each principal axis. The windward and leeward pressures in Tables 3.19 through 3.22 fall under case 1.

According to the exception in ASCE/SEI 27.3.5, building that meet the requirements of ASCE/SEI Appendix D.1 need only be designed for cases 1 and 3. It can be shown that this building is classified as a torsionally regular building under wind loads, which satisfies the requirements in ASCE/SEI D.4. Therefore, only cases 1 and 3 must be considered.

Surface(s)	Height above Ground Level, z, ft (m)	q, lb/ft² (kN/m²)	External Pressure qGC_p, lb/ft² (kN/m²)	Internal Pressure q_h(GC_{pi}), lb/ft² (kN/m²)	Net Pressure, p, lb/ft² (kN/m²)	
					+(GC_{pi})	−(GC_{pi})
1, 3, 5	All	64.4 (3.06)	−38.3 (−1.82)	±11.6 (±0.55)	−49.9 (−2.37)	−26.7 (−1.27)
2	52 (15.9)	64.4 (3.06)	43.8 (2.08)	±11.6 (±0.55)	32.2 (1.53)	55.4 (2.63)
	40 (12.2)	60.8 (2.89)	41.3 (1.97)	±11.6 (±0.55)	29.7 (1.42)	52.9 (2.52)
	30 (9.1)	57.3 (2.72)	39.0 (1.85)	±11.6 (±0.55)	27.4 (1.30)	50.6 (2.40)
	20 (6.1)	52.7 (2.50)	35.8 (1.70)	±11.6 (±0.55)	24.2 (1.15)	47.4 (2.25)
	10 (3.1)	49.7 (2.36)	33.8 (1.61)	±11.6 (±0.55)	22.2 (1.06)	45.4 (2.16)
4, 6	All	64.4 (3.06)	−27.4 (−1.30)	±11.6 (±0.55)	−39.0 (−1.85)	−15.8 (−0.75)
7, 8	—[1]	64.4 (3.06)	−49.2 (−2.34)	±11.6 (±0.55)	−60.8 (−2.89)	−37.6 (−1.79)
	—[2]	64.4 (3.06)	−49.2 (−2.34)	±11.6 (±0.55)	−60.8 (−2.89)	−37.6 (−1.79)
	—[3]	64.4 (3.06)	−27.4 (−1.30)	±11.6 (±0.55)	−39.0 (−1.85)	−15.8 (−0.75)
9	—	64.4 (3.06)	−68.9 (−3.28)	±11.6 (±0.55)	−80.5 (−3.83)	−57.3 (−2.73)
10	—	64.4 (3.06)	−37.2 (−1.77)	±11.6 (±0.55)	−48.8 (−2.32)	−25.6 (−1.22)
11	—	64.4 (3.06)	−49.2 (−2.34)	±11.6 (±0.55)	−60.8 (−2.89)	−37.6 (−1.79)

[1] From windward edge of roof to 26 ft (7.9 m).
[2] From 26 ft (7.9 m) to 52 ft (15.9 m).
[3] From 52 ft (15.9 m) to 55 ft (16.8 m).

TABLE 3.19 Design Wind Pressures, p, for Wind from North to South in Example 3.5

Surface(s)	Height above Ground Level, z, ft (m)	q, lb/ft² (kN/m²)	External Pressure qGC_p, lb/ft² (kN/m²)	Internal Pressure q_h(GC_{pi}), lb/ft² (kN/m²)	Net Pressure, p, lb/ft² (kN/m²) +(GC_{pi})	−(GC_{pi})
1, 3, 5	All	64.4 (3.06)	−38.3 (−1.82)	±11.6 (±0.55)	−49.9 (−2.37)	−26.7 (−1.27)
2	All	64.4 (3.06)	−27.4 (−1.30)	±11.6 (±0.55)	−39.0 (−1.85)	−15.8 (−0.75)
4, 6	52 (15.9)	64.4 (3.06)	43.8 (2.08)	±11.6 (±0.55)	32.2 (1.53)	55.4 (2.63)
	40 (12.2)	60.8 (2.89)	41.3 (1.97)	±11.6 (±0.55)	29.7 (1.42)	52.9 (2.52)
	30 (9.1)	57.3 (2.72)	39.0 (1.85)	±11.6 (±0.55)	27.4 (1.30)	50.6 (2.40)
	20 (6.1)	52.7 (2.50)	35.8 (1.70)	±11.6 (±0.55)	24.2 (1.15)	47.4 (2.25)
	10 (3.1)	49.7 (2.36)	33.8 (1.61)	±11.6 (±0.55)	22.2 (1.06)	45.4 (2.16)
7, 8	___(1)	64.4 (3.06)	−49.2 (−2.34)	±11.6 (±0.55)	−60.8 (−2.89)	−37.6 (−1.79)
	___(2)	64.4 (3.06)	−49.2 (−2.34)	±11.6 (±0.55)	−60.8 (−2.89)	−37.6 (−1.79)
	___(3)	64.4 (3.06)	−27.4 (−1.30)	±11.6 (±0.55)	−39.0 (−1.85)	−15.8 (−0.75)
9	—	64.4 (3.06)	−37.2 (−1.77)	±11.6 (±0.55)	−48.8 (−2.32)	−25.6 (−1.22)
10	—	64.4 (3.06)	−68.9 (−3.28)	±11.6 (±0.55)	−80.5 (−3.83)	−57.3 (−2.73)
11	—	64.4 (3.06)	−49.2 (−2.34)	±11.6 (±0.55)	−60.8 (−2.89)	−37.6 (−1.79)

(1) From windward edge of roof to 26 ft (7.9 m).
(2) From 26 ft (7.9 m) to 52 ft (15.9 m).
(3) From 52 ft (15.9 m) to 55 ft (16.8 m).

TABLE 3.20   Design Wind Pressures, p, for Wind from South to North in Example 3.5

Surface(s)	Height above Ground Level, z, ft (m)	q, lb/ft² (kN/m²)	External Pressure qGC_p, lb/ft² (kN/m²)	Internal Pressure $q_h(GC_{pi})$, lb/ft² (kN/m²)	Net Pressure, p, lb/ft² (kN/m²) $+(GC_{pi})$	Net Pressure, p, lb/ft² (kN/m²) $-(GC_{pi})$
1	All	64.4 (3.06)	−10.9 (−0.52)	±11.6 (±0.55)	−22.5 (−1.07)	0.7 (0.03)
2, 4, 6	All	64.4 (3.06)	−38.3 (−1.82)	±11.6 (±0.55)	−49.9 (−2.37)	−26.7 (−1.27)
3, 5	52 (15.9)	64.4 (3.06)	43.8 (2.08)	±11.6 (±0.55)	32.2 (1.53)	55.4 (2.63)
	40 (12.2)	60.8 (2.89)	41.3 (1.97)	±11.6 (±0.55)	29.7 (1.42)	52.9 (2.52)
	30 (9.1)	57.3 (2.72)	39.0 (1.85)	±11.6 (±0.55)	27.4 (1.30)	50.6 (2.40)
	20 (6.1)	52.7 (2.50)	35.8 (1.70)	±11.6 (±0.55)	24.2 (1.15)	47.4 (2.25)
	10 (3.1)	49.7 (2.36)	33.8 (1.61)	±11.6 (±0.55)	22.2 (1.06)	45.4 (2.16)
7	—	64.4 (3.06)	−16.4 (−0.78)	±11.6 (±0.55)	−28.0 (−1.33)	−4.8 (−0.23)
8, 11	—	64.4 (3.06)	−38.3 (−1.82)	±11.6 (±0.55)	−49.9 (−2.37)	−26.7 (−1.27)
9, 10	—[1]	64.4 (3.06)	−49.2 (−2.34)	±11.6 (±0.55)	−60.8 (−2.89)	−37.6 (−1.79)
	—[2]	64.4 (3.06)	−49.2 (−2.34)	±11.6 (±0.55)	−60.8 (−2.89)	−37.6 (−1.79)
	—[3]	64.4 (3.06)	−27.4 (−1.30)	±11.6 (±0.55)	−39.0 (−1.85)	−15.8 (−0.75)
	—[4]	64.4 (3.06)	−16.4 (−0.78)	±11.6 (±0.55)	−28.0 (−1.33)	−4.8 (−0.23)

[1] From windward edge of roof to 26 ft (7.9 m).
[2] From 26 ft (7.9 m) to 52 ft (15.9 m).
[3] From 52 ft (15.9 m) to 104 ft (31.7 m).
[4] From 104 ft (31.7 m) to 225 ft (68.6 m).

TABLE 3.21   Design Wind Pressures, p, for Wind from East to West in Example 3.5

Surface(s)	Height above Ground Level, z, ft (m)	q, lb/ft² (kN/m²)	External Pressure qGC_p, lb/ft² (kN/m²)	Internal Pressure q_h(GC_pi), lb/ft² (kN/m²)	Net Pressure, p, lb/ft² (kN/m²)	
					+(GC_pi)	−(GC_pi)
1	52 (15.9)	64.4 (3.06)	43.8 (2.08)	±11.6 (±0.55)	32.2 (1.53)	55.4 (2.63)
	40 (12.2)	60.8 (2.89)	41.3 (1.97)	±11.6 (±0.55)	29.7 (1.42)	52.9 (2.52)
	30 (9.1)	57.3 (2.72)	39.0 (1.85)	±11.6 (±0.55)	27.4 (1.30)	50.6 (2.40)
	20 (6.1)	52.7 (2.50)	35.8 (1.70)	±11.6 (±0.55)	24.2 (1.15)	47.4 (2.25)
	10 (3.1)	49.7 (2.36)	33.8 (1.61)	±11.6 (±0.55)	22.2 (1.06)	45.4 (2.16)
2, 4, 6	All	64.4 (3.06)	−38.3 (−1.82)	±11.6 (±0.55)	−49.9 (−2.37)	−26.7 (−1.27)
3, 5	All	64.4 (3.06)	−10.9 (−0.52)	±11.6 (±0.55)	−22.5 (−1.07)	0.7 (0.03)
7	—	64.4 (3.06)	−38.3 (−1.82)	±11.6 (±0.55)	−49.9 (−2.37)	−26.7 (−1.27)
8, 11	—	64.4 (3.06)	−16.4 (−0.78)	±11.6 (±0.55)	−28.0 (−1.33)	−4.8 (−0.23)
9, 10	—[1]	64.4 (3.06)	−49.2 (−2.34)	±11.6 (±0.55)	−60.8 (−2.89)	−37.6 (−1.79)
	—[2]	64.4 (3.06)	−49.2 (−2.34)	±11.6 (±0.55)	−60.8 (−2.89)	−37.6 (−1.79)
	—[3]	64.4 (3.06)	−27.4 (−1.30)	±11.6 (±0.55)	−39.0 (−1.85)	−15.8 (−0.75)
	—[4]	64.4 (3.06)	−16.4 (−0.78)	±11.6 (±0.55)	−28.0 (−1.33)	−4.8 (−0.23)

[1] From windward edge of roof to 26 ft (7.9 m).
[2] From 26 ft (7.9 m) to 52 ft (15.9 m).
[3] From 52 ft (15.9 m) to 104 ft (31.7 m).
[4] From 104 ft (31.7 m) to 225 ft (68.6 m).

TABLE 3.22 Design Wind Pressures, p, for Wind from West to East in Example 3.5

In case 3, 75 percent of the windward and leeward wall pressures act simultaneously on the building at each level above ground. Because the building is not symmetric, four combinations of wind loads must be considered for case 3 at each level. The combinations at the mean roof height are given in Fig. 3.25 for case 3.

The minimum design wind loading prescribed in ASCE/SEI 27.1.5 must be considered as a load case in addition to the load cases above.

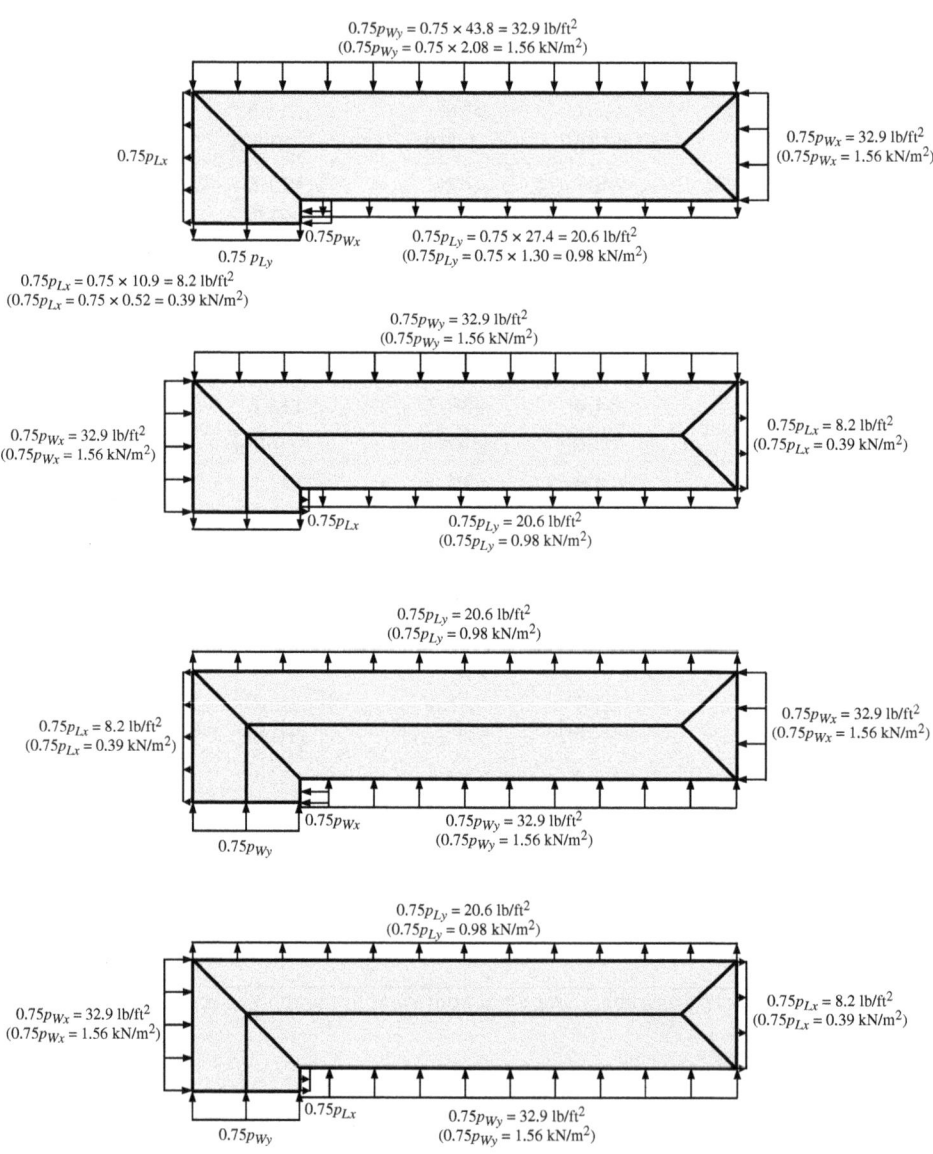

FIGURE 3.25 Load combinations at the mean roof height for case 3 for the building in Example 3.5.

3.4.6 Example 3.6—Wind Pressures on a Residential Building, MWFRS, Chapter 27, Part 1

Determine the wind pressures in both directions on the MWFRS of the residential building in Fig. 3.26 using the requirements in Part 1 of Chapter 27 and the design data in Table 3.23.

Solution
Check if the building meets all the conditions in ASCE/SEI 27.1.2, so that Part 1 in Chapter 27 can be used to determine the design wind pressures on the MWFRS.

The building is regular-shaped and does not have any unusual geometric irregularities in spatial form. Also, the building does not have any response characteristics that make it subject to across-wind loading or similar effects, and it is not sited at a location where channeling effects or buffeting in the wake of upwind obstructions need to be considered.

Therefore, the conditions in ASCE/SEI 27.1.2 are met and the requirements in Part 1 of Chapter 27 may be used.

The design procedure in Fig. 3.3 is used to determine the design wind pressures, p, in both directions.

Step 1—Determine the velocity pressure, q_z, over the height of the windward wall
Fig. 2.3

The flowchart in Fig. 2.3 is used to determine q_z.

- Step 1a—Determine the surface roughness category   ASCE/SEI 26.7.2

  In Table 3.23, the surface roughness is given as B.

- Step 1b—Determine the exposure category   ASCE/SEI 26.7.3

  It is assumed that surface roughness B applies in all directions and that Exposures C and D are not applicable. Therefore, the exposure category is B.

- Step 1c—Determine the terrain exposure constants   ASCE/SEI Table 26.11-1

  For Exposure B, $\alpha = 7.0$ and $z_g = 1,200$ ft (365.76 m).

- Step 1d—Determine the velocity pressure exposure coefficient, K_z
       ASCE/SEI Table 26.10-1

  Values of K_z are given in Table 3.24 for Exposure B.

  For example, at $z = 152$ ft (46.3 m):

  $$K_z = 2.01(z/z_g)^{2/\alpha} = 2.01 \times (152/1,200)^{2/7.0} = 1.11$$

  In S.I.:

  $$K_z = 2.01 \times (46.3/365.76)^{2/7.0} = 1.11$$

- Step 1e—Determine the topographic factors, K_{zt} ASCE/SEI 26.8

  Because the building is not located on a hill, ridge, or escarpment, $K_{zt} = 1.0$.

- Step 1f—Determine the wind directionality factor, K_d ASCE/SEI Table 26.6-1

  For the MWFRS of a building structure, $K_d = 0.85$.

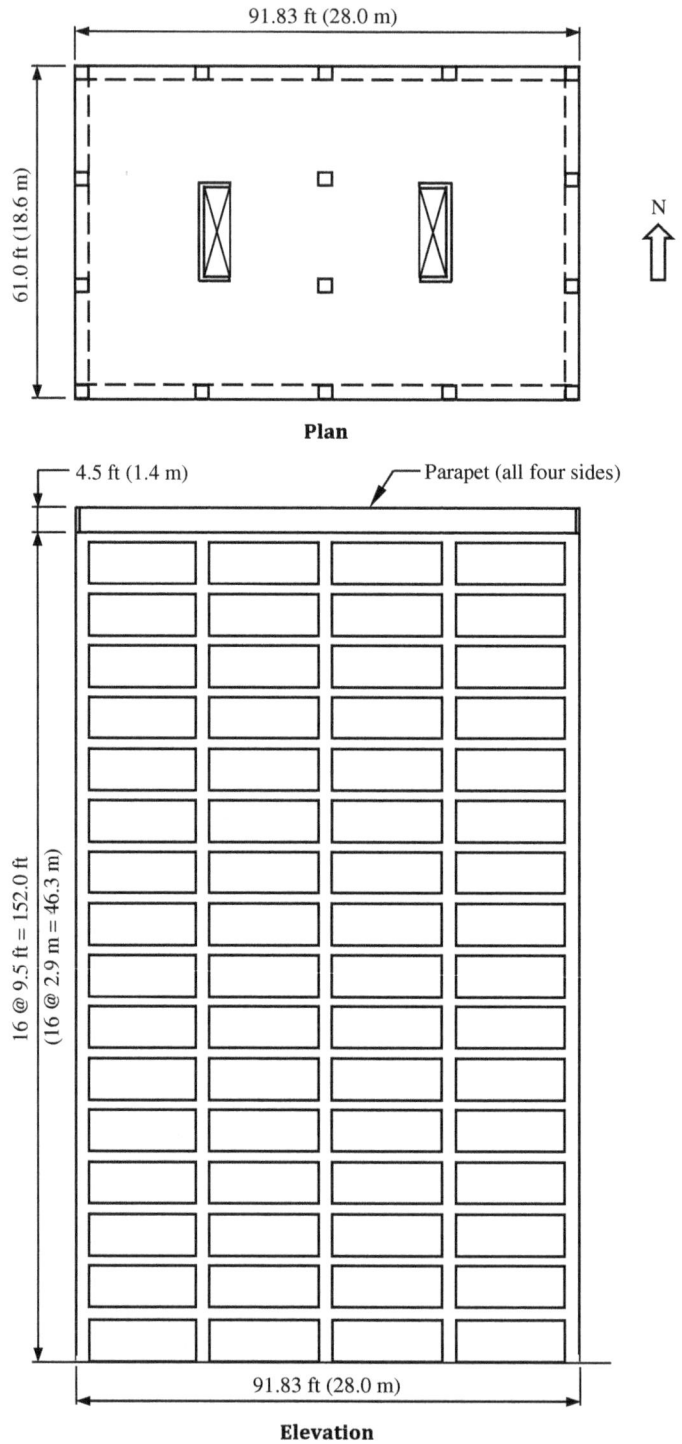

FIGURE 3.26   Plan and elevation of the residential building in Example 3.6.

Location	Atlanta, GA
Surface roughness	B
Topography	Not situated on a hill, ridge, or escarpment
Occupancy	Less than 300 people congregate in one area at the same time
Enclosure classification	Enclosed
MWFRS	Combination of reinforced concrete shear walls and moment frames

TABLE 3.23   Design Data for the Residential Building in Example 3.6

- Step 1g—Determine the ground elevation factor, K_e ASCE/SEI 26.9
  Ground elevation factor can be taken as 1.0 for all elevations.
- Step 1h—Determine the risk category of the building   ASCE/SEI Table 1.5-1
  Due to the nature of its occupancy, this residential building falls under Risk Category II.
- Step 1i—Determine the basic wind speed, V Table 2.1
  For Risk Category II, use IBC Figure 1609.3(1) or ASCE/SEI Figure 26.5-1B. Equivalently, use Ref. 3 or Ref. 4 to obtain $V = 107$ mi/h (48 m/s) for Atlanta, GA.

Height above Ground Level, z, ft (m)	K_z
152.0 (46.3)	1.11
142.5 (43.4)	1.09
133.0 (40.5)	1.07
123.5 (37.6)	1.05
114.0 (34.8)	1.03
104.5 (31.9)	1.00
95.0 (29.0)	0.97
85.5 (26.1)	0.95
76.0 (23.2)	0.91
66.5 (20.3)	0.88
57.0 (17.4)	0.84
47.5 (14.5)	0.80
38.0 (11.6)	0.75
28.5 (8.7)	0.69
19.0 (5.8)	0.62
9.5 (2.9)	0.57

TABLE 3.24   Velocity Pressure Exposure Coefficient, K_z, for the Building in Example 3.6

- Step 1j—Determine the wind velocity pressure, q_z ASCE/SEI 26.10.2

$$q_z = 0.00256 K_z K_{zt} K_d K_e V^2$$

$$= 0.00256 \times K_z \times 1.0 \times 0.85 \times 1.0 \times 107^2 = 24.9 K_z \text{ lb/ft}^2$$

In S.I.:

$$q_z = 0.613 K_z K_{zt} K_d K_e V^2$$

$$= 0.613 \times K_z \times 1.0 \times 0.85 \times 1.0 \times 48^2 / 1,000 = 1.20 K_z \text{ kN/m}^2$$

The velocity pressures over the height of the windward wall are given in Table 3.25.

Step 2—Determine the velocity pressure, q_h, on the roof and side walls   *Fig. 2.3*

From Step 1, $q_h = 27.6$ lb/ft² (1.33 kN/m²).

Step 3—Determine the gust-effect factor   *Fig. 2.4*

- Step 3a—Determine the fundamental natural frequency, n_1

In lieu of obtaining n_1 from a dynamic analysis of the building, check if the approximate lower bound natural frequency, n_a, given in ASCE/SEI 26.11.3 can be used for this building:

Height above Ground Level, z, ft (m)	K_z	q_z, lb/ft² (kN/m²)
152.0 (46.3)	1.11	27.6 (1.33)
142.5 (43.4)	1.09	27.1 (1.31)
133.0 (40.5)	1.07	26.6 (1.28)
123.5 (37.6)	1.05	26.2 (1.26)
114.0 (34.8)	1.03	25.7 (1.24)
104.5 (31.9)	1.00	24.9 (1.20)
95.0 (29.0)	0.97	24.2 (1.16)
85.5 (26.1)	0.95	23.7 (1.14)
76.0 (23.2)	0.91	22.7 (1.09)
66.5 (20.3)	0.88	21.9 (1.06)
57.0 (17.4)	0.84	20.9 (1.01)
47.5 (14.5)	0.80	19.9 (0.96)
38.0 (11.6)	0.75	18.7 (0.90)
28.5 (8.7)	0.69	17.2 (0.83)
19.0 (5.8)	0.62	15.4 (0.74)
9.5 (2.9)	0.57	14.2 (0.68)

TABLE 3.25   Velocity Pressure, q_z, for the Building in Example 3.6

- Building height = 152 ft (46.3 m) < 300 ft (91.4 m)
- Building height = 152 ft (46.3 m) < $4L_{eff} = 4 \times 61 = 244$ ft (74.4 m) in the N-S direction

$$= 152 \text{ ft (46.3 m)} < 4L_{eff} = 4 \times 91.83 = 367.3 \text{ ft (112.0 m) in the E-W direction}$$

Because both of these limitations are satisfied, ASCE/SEI Equation (26.11-5) can be used to determine n_a in both directions for the MWFRS consisting of reinforced concrete shear walls and moment frames.

For a concrete building with a MWFRS other than moment-resisting frames, n_a can be determined by the following:

$$n_a = \frac{75}{h} = \frac{75}{152} = 0.49 \text{ Hz} < 1.0 \text{ Hz} \qquad \text{ASCE/SEI Equation (26.11-4)}$$

In S.I.:

$$n_a = \frac{22.86}{h} = \frac{22.86}{46.3} = 0.49 \text{ Hz} < 1.0 \text{ Hz}$$

Therefore, the building is flexible in both directions.

- Step 3b—Determine the gust-effect factor, G_f

  For flexible buildings, G_f is determined by ASCE/SEI Equation (26.11-10).

  Calculations for G_f in both principal directions are given in Table 3.26.

Step 4—Determine the external pressure coefficients, C_p   *ASCE/SEI Figure 27.3-1*

For a flat roof, assume a ridge line occurs parallel to the long dimension of the building. In this example, the ridge line is placed in the E-W direction.

- N-S wind

  Windward wall: $C_p = 0.8$

  Leeward wall: $L/B = 61/91.83 = 0.66$; $C_p = -0.5$

  Side walls: $C_p = -0.7$

  Roof: normal to the ridge with $\theta < 10$ degrees and $h/L = 152/61 = 2.5$

  From 0 to 61 ft (0 to 18.6 m): $C_p = -1.3, -0.18$

  The value of -1.3 is permitted to be reduced linearly with the area over which it is applicable in accordance with footnote b in ASCE/SEI Figure 27.3-1.

  Area over which -1.3 is applicable = $61 \times 91.83 = 5,602$ ft² (520.4 m²) > 1,000 ft² (92.9 m²)

  Therefore, $C_p = 0.8 \times (-1.3) = -1.04$.

- E-W wind

  Windward wall: $C_p = 0.8$

  Leeward wall: $L/B = 91.83/61 = 1.5$; $C_p = \dfrac{-0.5 - 0.3}{2} = -0.4$

  Side walls: $C_p = -0.7$

  Roof: parallel to ridge with $h/L = 152/91.83 = 1.7$

Calculations	ASCE/SEI Reference
$g_Q = g_v = 3.4$	26.11.5
$g_R = \sqrt{2\ln(3{,}600 n_1)} + [0.577/\sqrt{2\ln(3{,}600 n_1)}] = 4.0$	Equation (26.11-11)
$z_{min} = 30.0$ ft (9.14 m)	Table 26.11-1
$\bar{z} = \text{greater of} \begin{cases} 0.6h = 91.2 \text{ ft } (27.80 \text{ m}) \text{ [governs]} \\ z_{min} = 30.0 \text{ ft } (9.14 \text{ m}) \end{cases}$	26.11.4
$\ell = 320.0$ ft (97.54 m)	Table 26.11-1
$c = 0.30$	Table 26.11-1
$\bar{\varepsilon} = 1/3.0$	Table 26.11-1
$I_{\bar{z}} = c(33/\bar{z})^{1/6} = 0.25$ [In S.I.: $c(10/\bar{z})^{1/6} = 0.25$]	Equation (26.11-7)
$L_{\bar{z}} = \ell(\bar{z}/33)^{\bar{\varepsilon}} = 449.1$ ft [In S.I.: $\ell(\bar{z}/10)^{\bar{\varepsilon}} = 137.15$ m]	Equation (26.11-9)
$Q = \sqrt{\dfrac{1}{1 + 0.63\left(\dfrac{B+h}{L_{\bar{z}}}\right)^{0.63}}} = \begin{cases} 0.84 \text{ in the N-S direction with } B = 91.83 \text{ ft } (27.99 \text{ m}) \\ 0.85 \text{ in the E-W direction with } B = 61.0 \text{ ft } (18.59 \text{ m}) \end{cases}$	Equation (26.11-8)
Damping ratio $\beta = 0.015$ for concrete buildings	C26.11
$\bar{b} = 0.45$	Table 26.11-1
$\bar{\alpha} = 1/4.0$	Table 26.11-1
$\bar{V}_{\bar{z}} = \bar{b}\left(\dfrac{\bar{z}}{33}\right)^{\bar{\alpha}}\left(\dfrac{88}{60}\right)V = 91.1$ ft/s $\left[\text{In S.I.: } \bar{V}_{\bar{z}} = \bar{b}\left(\dfrac{\bar{z}}{10}\right)^{\bar{\alpha}}V = 27.89 \text{ m/s}\right]$	Equation (26.11-16)
$N_1 = n_1 L_{\bar{z}}/\bar{V}_{\bar{z}} = 2.42$	Equation (26.11-14)
$R_n = \dfrac{7.47N_1}{(1 + 10.3N_1)^{5/3}} = 0.08$	Equation (26.11-13)
$\eta_h = 4.6 n_1 h/\bar{V}_{\bar{z}} = 3.8$	26.11.5
$R_h = \dfrac{1}{\eta_h} - \dfrac{1}{2\eta_h^2}(1 - e^{-2\eta_h}) = 0.23$	Equation (26.11-15a)
$\eta_B = 4.6 n_1 B/\bar{V}_{\bar{z}} = \begin{cases} 2.3 \text{ in the N-S direction with } B = 91.83 \text{ ft } (27.99 \text{ m}) \\ 1.5 \text{ in the E-W direction with } B = 61.0 \text{ ft } (18.59 \text{ m}) \end{cases}$	26.11.5
$R_B = \dfrac{1}{\eta_B} - \dfrac{1}{2\eta_B^2}(1 - e^{-2\eta_B}) = \begin{cases} 0.34 \text{ in the N-S direction} \\ 0.46 \text{ in the E-W direction} \end{cases}$	Equation (26.11-15a)
$\eta_L = 15.4 n_1 L/\bar{V}_{\bar{z}} = \begin{cases} 5.1 \text{ in the N-S direction with } L = 61.0 \text{ ft } (18.59 \text{ m}) \\ 7.6 \text{ in the E-W direction with } L = 91.83 \text{ ft } (27.99 \text{ m}) \end{cases}$	26.11.5

TABLE 3.26 Determination of Gust-Effect Factor, G_f, for the Building in Example 3.6

Calculations		ASCE/SEI Reference
$R_L = \dfrac{1}{\eta_L} - \dfrac{1}{2\eta_L^2}(1 - e^{-2\eta_L}) = \begin{cases} \end{cases}$	0.18 in the N-S direction 0.12 in the E-W direction	Equation (26.11-15a)
$R = \sqrt{R_n R_h R_B (0.53 + 0.47 R_L)/\beta} = \begin{cases} \end{cases}$	0.51 in the N-S direction 0.58 in the E-W direction	Equation (26.11-12)
$G_f = 0.925 \left(\dfrac{1 + 1.7 I_{\bar{z}} \sqrt{g_Q^2 Q^2 + g_R^2 R^2}}{1 + 1.7 g_v I_{\bar{z}}} \right) = \begin{cases} \end{cases}$	0.94 in the N-S direction 0.97 in the E-W direction	Equation (26.11-10)

TABLE 3.26 Determination of Gust-Effect Factor, G_f, for the Building in Example 3.6 (*Continued*)

From 0 to 76 ft (0 to 23.2 m): $C_p = -1.3, -0.18$

Area over which −1.3 is applicable = $76 \times 61 = 4{,}636$ ft^2 (430.7 m^2) > 1,000 ft^2 (92.9 m^2)

Therefore, $C_p = 0.8 \times (-1.3) = -1.04$.

From 76 ft to 91.83 ft (23.2 m to 28.0 m): $C_p = -0.7, -0.18$

The roof pressure coefficient $C_p = -0.18$ may become critical where wind loads are combined with roof live loads or snow loads. Determination of wind pressures based on this pressure coefficient should be performed, but such calculations are not shown in this example.

Step 5—Determine the velocity pressure for internal pressure determination, q_i   Sec. 3.2.1

   According to ASCE/SEI 27.3.1, $q_i = q_h = 27.6$ lb/ft^2 (1.33 kN/m^2)

Step 6—Determine the enclosure classification   Table 2.7

   In the design data, the building is given as enclosed.

Step 7—Determine the internal pressure coefficient, (GC_{pi})   Table 2.9

   For an enclosed building, $(GC_{pi}) = +0.18, -0.18$.

Step 8—Determine the design wind pressures, p

- Windward walls for wind in the N-S direction:

$$p_z = q_z G_f C_p - q_h (GC_{pi})$$

$$= (q_z \times 0.94 \times 0.8) - [27.6 \times (\pm 0.18)]$$

$$= (0.75 q_z \mp 5.0) \text{ lb/ft}^2$$

In S.I.:

$$p_z = (q_z \times 0.94 \times 0.8) - [1.33 \times (\pm 0.18)] = (0.75q_z \mp 0.24) \text{ kN/m}^2$$

- Windward walls for wind in the E-W direction:

$$p_z = q_z G_f C_p - q_h (GC_{pi})$$
$$= (q_z \times 0.97 \times 0.8) - [27.6 \times (\pm 0.18)]$$
$$= (0.78q_z \mp 5.0) \text{ lb/ft}^2$$

In S.I.:

$$p_z = (q_z \times 0.97 \times 0.8) - [1.33 \times (\pm 0.18)] = (0.78q_z \mp 0.24) \text{ kN/m}^2$$

- Leeward wall, side walls, and roof for wind in the N-S direction

$$p_h = q_h G_f C_p - q_h (GC_{pi})$$
$$= (27.6 \times 0.94 \times C_p) - [27.6 \times (\pm 0.18)]$$
$$= (25.9C_p \mp 5.0) \text{ lb/ft}^2$$

In S.I.:

$$p_h = (1.33 \times 0.94 \times C_p) - [1.33 \times (\pm 0.18)] = (1.25C_p \mp 0.24) \text{ kN/m}^2$$

- Leeward wall, side walls, and roof for wind in the E-W direction

$$p_h = q_h G_f C_p - q_h (GC_{pi})$$
$$= (27.6 \times 0.97 \times C_p) - [27.6 \times (\pm 0.18)]$$
$$= (26.8C_p \mp 5.0) \text{ lb/ft}^2$$

In S.I.:

$$p_h = (1.33 \times 0.97 \times C_p) - [1.33 \times (\pm 0.18)] = (1.29C_p \mp 0.24) \text{ kN/m}^2$$

Design wind pressures for wind in the N-S and E-W directions are given in Tables 3.27 and 3.28, respectively.

- Design wind pressure on the parapet

$$q_z = 0.00256 K_z K_{zt} K_d K_e V^2$$

In S.I.:

$$q_z = 0.613 K_z K_{zt} K_d K_e V^2$$

Top of parapet:

$$z = 152 + 4.5 = 156.5 \text{ ft (47.7 m)}$$

$$K_z = 2.01(z/z_g)^{2/\alpha} = 2.01 \times (156.5/1,200)^{2/7.0} = 1.12$$

Building Surface	Height above Ground Level, z, ft (m)	q, lb/ft² (kN/m²)	External Pressure qG_fC_p, lb/ft² (kN/m²)	Internal Pressure $q_h(GC_{pi})$, lb/ft² (kN/m²)	Net Pressure, p, lb/ft² (kN/m²)	
					$+(GC_{pi})$	$-(GC_{pi})$
Windward wall	152.0 (46.3)	27.6 (1.33)	20.7 (1.00)	±5.0 (±0.24)	15.7 (0.76)	25.7 (1.24)
	142.5 (43.4)	27.1 (1.31)	20.3 (0.98)	±5.0 (±0.24)	15.3 (0.74)	25.3 (1.22)
	133.0 (40.5)	26.6 (1.28)	20.0 (0.96)	±5.0 (±0.24)	15.0 (0.72)	25.0 (1.20)
	123.5 (37.6)	26.2 (1.26)	19.7 (0.95)	±5.0 (±0.24)	14.7 (0.71)	24.7 (1.19)
	114.0 (34.8)	25.7 (1.24)	19.3 (0.93)	±5.0 (±0.24)	14.3 (0.69)	24.3 (1.17)
	104.5 (31.9)	24.9 (1.20)	18.7 (0.90)	±5.0 (±0.24)	13.7 (0.66)	23.7 (1.14)
	95.0 (29.0)	24.2 (1.16)	18.2 (0.87)	±5.0 (±0.24)	13.2 (0.63)	23.2 (1.11)
	85.5 (26.1)	23.7 (1.14)	17.8 (0.86)	±5.0 (±0.24)	12.8 (0.62)	22.8 (1.10)
	76.0 (23.2)	22.7 (1.09)	17.0 (0.82)	±5.0 (±0.24)	12.0 (0.58)	22.0 (1.06)
	66.5 (20.3)	21.9 (1.06)	16.4 (0.80)	±5.0 (±0.24)	11.4 (0.56)	21.4 (1.04)
	57.0 (17.4)	20.9 (1.01)	15.7 (0.76)	±5.0 (±0.24)	10.7 (0.52)	20.7 (1.00)
	47.5 (14.5)	19.9 (0.96)	14.9 (0.72)	±5.0 (±0.24)	9.9 (0.48)	19.9 (0.96)
	38.0 (11.6)	18.7 (0.90)	14.0 (0.68)	±5.0 (±0.24)	9.0 (0.44)	19.0 (0.92)
	28.5 (8.7)	17.2 (0.83)	12.9 (0.62)	±5.0 (±0.24)	7.9 (0.38)	17.9 (0.86)
	19.0 (5.8)	15.4 (0.74)	11.6 (0.56)	±5.0 (±0.24)	6.6 (0.32)	16.6 (0.80)
	9.5 (2.9)	14.2 (0.68)	10.7 (0.51)	±5.0 (±0.24)	5.7 (0.27)	15.7 (0.75)
Leeward wall	All	27.6 (1.33)	−13.0 (−0.63)	±5.0 (±0.24)	−18.0 (−0.87)	−8.0 (−0.39)
Side walls	All	27.6 (1.33)	−18.1 (−0.88)	±5.0 (±0.24)	−23.1 (−1.12)	−13.1 (−0.64)
Roof	—[1]	27.6 (1.33)	−26.9 (−1.30)	±5.0 (±0.24)	−31.9 (−1.54)	−21.9 (−1.06)

[1] From windward edge of roof to 61.0 ft (18.6 m).

TABLE 3.27 Design Wind Pressures, p, for Wind in the N-S Direction for the Building in Example 3.6

Building Surface	Height above Ground Level, z, ft (m)	q, lb/ft² (kN/m²)	External Pressure qG_fC_p, lb/ft² (kN/m²)	Internal Pressure q_h(GC_{pi}), lb/ft² (kN/m²)	Net Pressure, p, lb/ft² (kN/m²) +(GC_{pi})	−(GC_{pi})
Windward wall	152.0 (46.3)	27.6 (1.33)	21.5 (1.04)	±5.0 (±0.24)	16.5 (0.80)	26.5 (1.28)
	142.5 (43.4)	27.1 (1.31)	21.1 (1.02)	±5.0 (±0.24)	16.1 (0.78)	26.1 (1.26)
	133.0 (40.5)	26.6 (1.28)	20.8 (1.00)	±5.0 (±0.24)	15.8 (0.76)	25.8 (1.24)
	123.5 (37.6)	26.2 (1.26)	20.4 (0.98)	±5.0 (±0.24)	15.4 (0.74)	25.4 (1.22)
	114.0 (34.8)	25.7 (1.24)	20.1 (0.97)	±5.0 (±0.24)	15.1 (0.73)	25.1 (1.21)
	104.5 (31.9)	24.9 (1.20)	19.4 (0.94)	±5.0 (±0.24)	14.4 (0.70)	24.4 (1.18)
	95.0 (29.0)	24.2 (1.16)	18.9 (0.91)	±5.0 (±0.24)	13.9 (0.67)	23.9 (1.15)
	85.5 (26.1)	23.7 (1.14)	18.5 (0.89)	±5.0 (±0.24)	13.5 (0.65)	23.5 (1.13)
	76.0 (23.2)	22.7 (1.09)	17.7 (0.85)	±5.0 (±0.24)	12.7 (0.61)	22.7 (1.09)
	66.5 (20.3)	21.9 (1.06)	17.1 (0.83)	±5.0 (±0.24)	12.1 (0.59)	22.1 (1.07)
	57.0 (17.4)	20.9 (1.01)	16.3 (0.79)	±5.0 (±0.24)	11.3 (0.55)	21.3 (1.03)
	47.5 (14.5)	19.9 (0.96)	15.5 (0.75)	±5.0 (±0.24)	10.5 (0.51)	20.5 (0.99)
	38.0 (11.6)	18.7 (0.90)	14.6 (0.70)	±5.0 (±0.24)	9.6 (0.46)	19.6 (0.94)
	28.5 (8.7)	17.2 (0.83)	13.4 (0.65)	±5.0 (±0.24)	8.4 (0.41)	18.4 (0.89)
	19.0 (5.8)	15.4 (0.74)	12.0 (0.58)	±5.0 (±0.24)	7.0 (0.34)	17.0 (0.82)
	9.5 (2.9)	14.2 (0.68)	11.1 (0.53)	±5.0 (±0.24)	6.1 (0.29)	16.1 (0.77)
Leeward wall	All	27.6 (1.33)	−10.7 (−0.52)	±5.0 (±0.24)	−15.7 (−0.76)	−5.7 (−0.28)
Side walls	All	27.6 (1.33)	−18.8 (−0.90)	±5.0 (±0.24)	−23.8 (−1.14)	−13.8 (−0.66)
Roof	___(1)	27.6 (1.33)	−27.9 (−1.34)	±5.0 (±0.24)	−32.9 (−1.58)	−22.9 (−1.10)
	___(2)	27.6 (1.33)	−18.8 (−0.90)	±5.0 (±0.24)	−23.8 (−1.14)	−13.8 (−0.66)

(1) From windward edge of roof to 76 ft (23.2 m).
(2) From 76 ft (23.2 m) to 91.83 ft (28.0 m).

TABLE 3.28   Design Wind Pressures, p, for Wind in the E-W Direction for the Building in Example 3.6

$$q_p = 0.00256 \times 1.12 \times 1.0 \times 0.85 \times 1.0 \times 107^2 = 27.9 \text{ lb/ft}^2$$

In S.I.:

$$K_z = 2.01 \times (47.7/365.76)^{2/7.0} = 1.12$$

$$q_p = 0.613 \times 1.12 \times 1.0 \times 0.85 \times 1.0 \times 48^2 / 1,000 = 1.35 \text{ kN/m}^2$$

$$p_p = q_p (GC_{pn}) \hspace{3cm} \text{ASCE/SEI Equation (27.3-3)}$$

Winward parapet:

$$p_p = 27.9 \times 1.5 = 41.9 \text{ lb/ft}^2 \quad (\text{in S.I.:} \quad p_p = 1.35 \times 1.5 = 2.03 \text{ kN/m}^2)$$

Leeward parapet:

$$p_p = 27.9 \times (-1.0) = -27.9 \text{ lb/ft}^2 \quad (\text{in S.I.:} \quad p_p = 1.35 \times (-1.0) = -1.35 \text{ kN/m}^2)$$

Net design wind pressures in the N-S and E-W directions for positive and negative internal pressures are given in Figs. 3.27 and 3.28, respectively.

The horizontal wind pressures on the MWFRS for both directions are shown in Fig. 3.29. It is evident from Figs. 3.27, 3.28, and 3.29 that the internal pressures cancel out when determining the horizontal wind pressures on the MWFRS.

The building must be designed for the wind load cases defined in ASCE/SEI Figure 27.3-8 (see Fig. 3.8).

In case 1, the full design wind pressures act on the projected area perpendicular to each principal axis of the building at each level above ground. These pressures act separately along each principal axis. The windward and leeward pressures in Fig. 3.29 fall under case 1.

In case 2, 75 percent of the design wind pressures on the windward and leeward walls are applied on the projected area perpendicular to each principal axis of the building along with a torsional moment. The wind pressures and torsional moments, which vary over the height of the building, are applied separately for each principal axis.

The torsional moment, M_T, is calculated using the eccentricity, e, for flexible buildings, which is determined by ASCE/SEI Equation (27.3-4):

$$e = \frac{e_Q + 1.7 I_{\bar{z}} \sqrt{(g_Q Q e_Q)^2 + (g_R R e_R)^2}}{1 + 1.7 I_{\bar{z}} \sqrt{(g_Q Q)^2 + (g_R R)^2}}$$

where e_Q = eccentricity e determined for rigid buildings in ASCE/SEI Figure 27.3-8

e_R = distance between the elastic shear center and the center of mass of each floor

The other items in this equation are given in Table 3.26.

Calculations for e are given in Table 3.29 for both principal wind directions. Because the locations of the elastic shear center and center of mass coincide in both principal directions, $e_R = 0$.

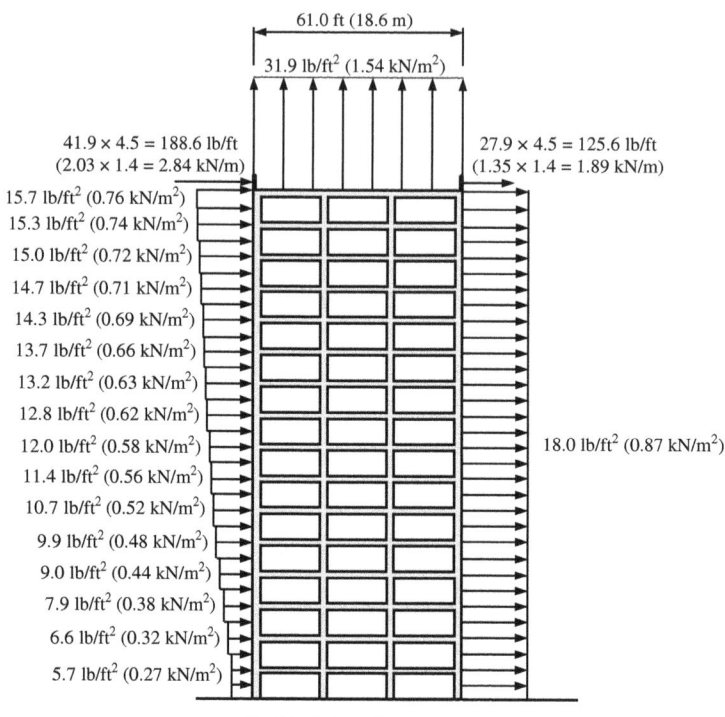

Positive internal pressure

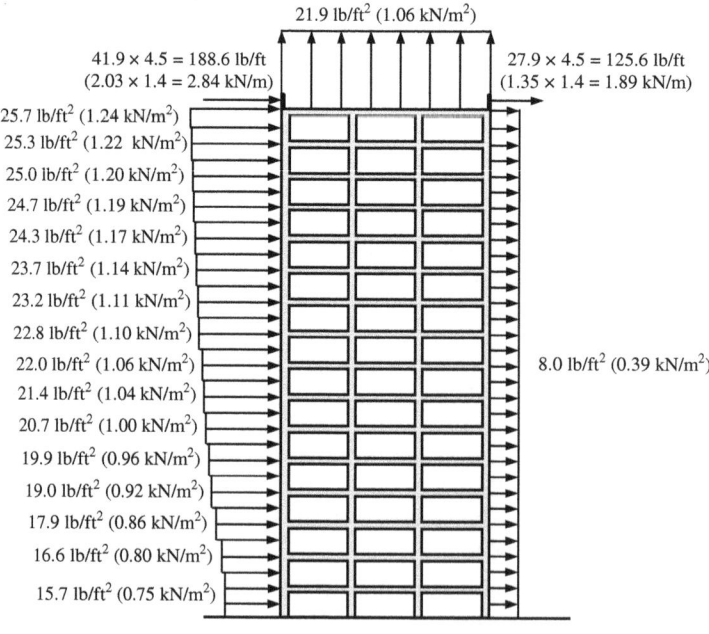

Negative internal pressure

FIGURE 3.27 Net design wind pressures in the N-S direction for the building in Example 3.6.

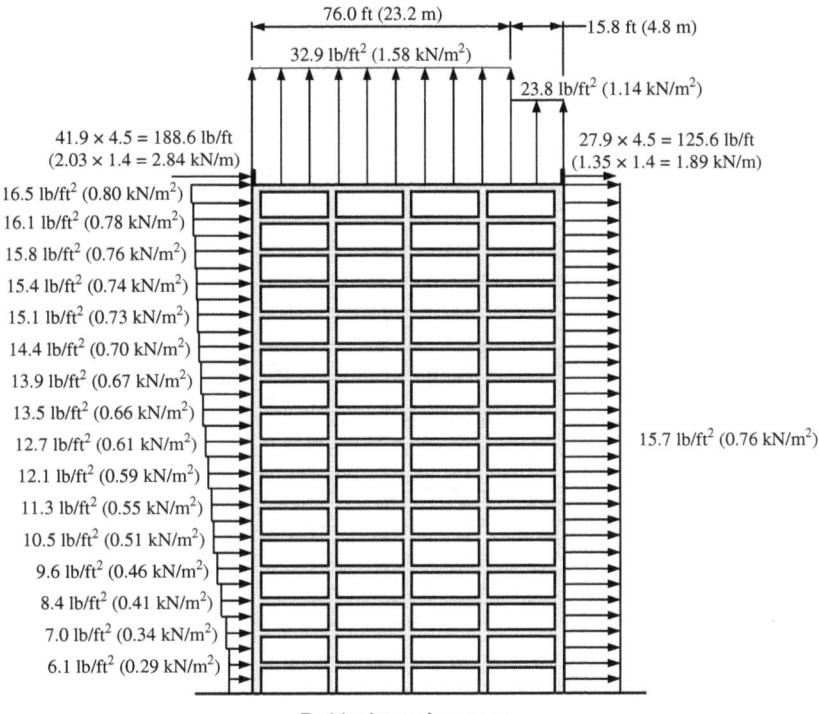

76.0 ft (23.2 m)   15.8 ft (4.8 m)

32.9 lb/ft² (1.58 kN/m²)

23.8 lb/ft² (1.14 kN/m²)

41.9 × 4.5 = 188.6 lb/ft
(2.03 × 1.4 = 2.84 kN/m)

27.9 × 4.5 = 125.6 lb/ft
(1.35 × 1.4 = 1.89 kN/m)

16.5 lb/ft² (0.80 kN/m²)
16.1 lb/ft² (0.78 kN/m²)
15.8 lb/ft² (0.76 kN/m²)
15.4 lb/ft² (0.74 kN/m²)
15.1 lb/ft² (0.73 kN/m²)
14.4 lb/ft² (0.70 kN/m²)
13.9 lb/ft² (0.67 kN/m²)
13.5 lb/ft² (0.66 kN/m²)
12.7 lb/ft² (0.61 kN/m²)
12.1 lb/ft² (0.59 kN/m²)
11.3 lb/ft² (0.55 kN/m²)
10.5 lb/ft² (0.51 kN/m²)
9.6 lb/ft² (0.46 kN/m²)
8.4 lb/ft² (0.41 kN/m²)
7.0 lb/ft² (0.34 kN/m²)
6.1 lb/ft² (0.29 kN/m²)

15.7 lb/ft² (0.76 kN/m²)

Positive internal pressure

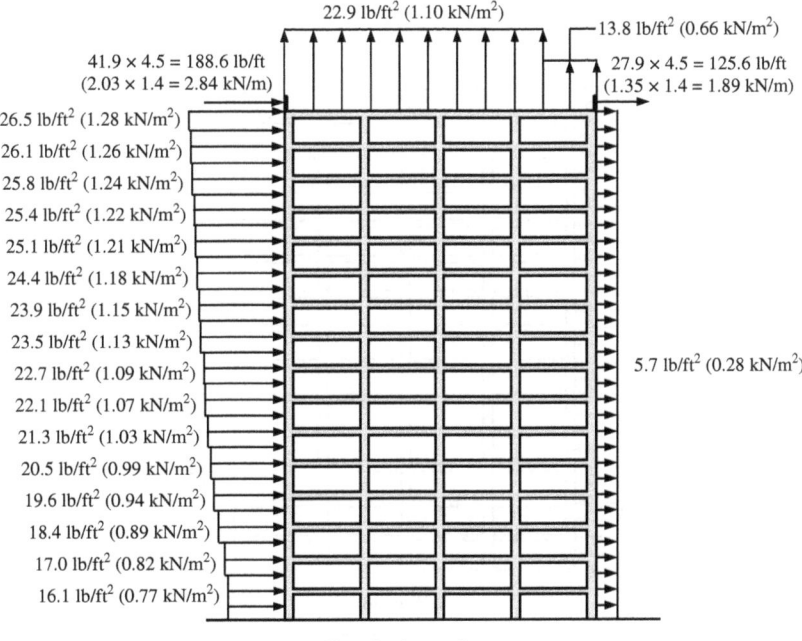

22.9 lb/ft² (1.10 kN/m²)

13.8 lb/ft² (0.66 kN/m²)

41.9 × 4.5 = 188.6 lb/ft
(2.03 × 1.4 = 2.84 kN/m)

27.9 × 4.5 = 125.6 lb/ft
(1.35 × 1.4 = 1.89 kN/m)

26.5 lb/ft² (1.28 kN/m²)
26.1 lb/ft² (1.26 kN/m²)
25.8 lb/ft² (1.24 kN/m²)
25.4 lb/ft² (1.22 kN/m²)
25.1 lb/ft² (1.21 kN/m²)
24.4 lb/ft² (1.18 kN/m²)
23.9 lb/ft² (1.15 kN/m²)
23.5 lb/ft² (1.13 kN/m²)
22.7 lb/ft² (1.09 kN/m²)
22.1 lb/ft² (1.07 kN/m²)
21.3 lb/ft² (1.03 kN/m²)
20.5 lb/ft² (0.99 kN/m²)
19.6 lb/ft² (0.94 kN/m²)
18.4 lb/ft² (0.89 kN/m²)
17.0 lb/ft² (0.82 kN/m²)
16.1 lb/ft² (0.77 kN/m²)

5.7 lb/ft² (0.28 kN/m²)

Negative internal pressure

FIGURE 3.28   Net design wind pressures in the E-W direction for the building in Example 3.6.

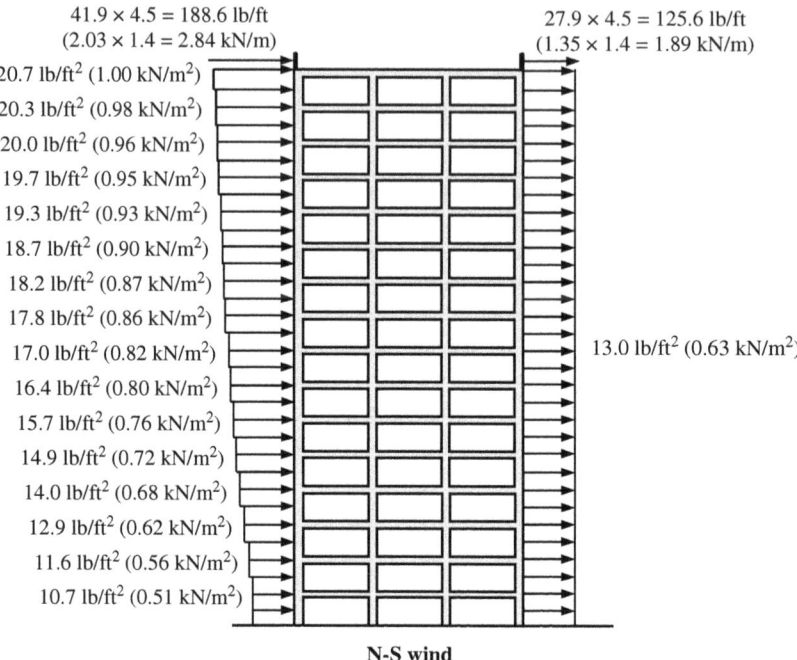

41.9 × 4.5 = 188.6 lb/ft
(2.03 × 1.4 = 2.84 kN/m)

27.9 × 4.5 = 125.6 lb/ft
(1.35 × 1.4 = 1.89 kN/m)

20.7 lb/ft² (1.00 kN/m²)
20.3 lb/ft² (0.98 kN/m²)
20.0 lb/ft² (0.96 kN/m²)
19.7 lb/ft² (0.95 kN/m²)
19.3 lb/ft² (0.93 kN/m²)
18.7 lb/ft² (0.90 kN/m²)
18.2 lb/ft² (0.87 kN/m²)
17.8 lb/ft² (0.86 kN/m²)
17.0 lb/ft² (0.82 kN/m²)
16.4 lb/ft² (0.80 kN/m²)
15.7 lb/ft² (0.76 kN/m²)
14.9 lb/ft² (0.72 kN/m²)
14.0 lb/ft² (0.68 kN/m²)
12.9 lb/ft² (0.62 kN/m²)
11.6 lb/ft² (0.56 kN/m²)
10.7 lb/ft² (0.51 kN/m²)

13.0 lb/ft² (0.63 kN/m²)

N-S wind

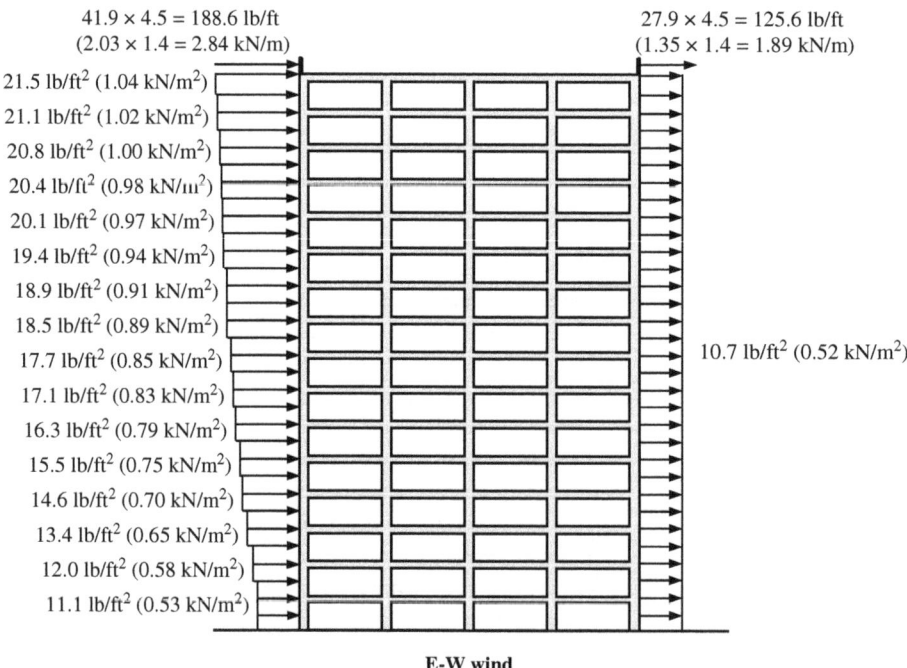

41.9 × 4.5 = 188.6 lb/ft
(2.03 × 1.4 = 2.84 kN/m)

27.9 × 4.5 = 125.6 lb/ft
(1.35 × 1.4 = 1.89 kN/m)

21.5 lb/ft² (1.04 kN/m²)
21.1 lb/ft² (1.02 kN/m²)
20.8 lb/ft² (1.00 kN/m²)
20.4 lb/ft² (0.98 kN/m²)
20.1 lb/ft² (0.97 kN/m²)
19.4 lb/ft² (0.94 kN/m²)
18.9 lb/ft² (0.91 kN/m²)
18.5 lb/ft² (0.89 kN/m²)
17.7 lb/ft² (0.85 kN/m²)
17.1 lb/ft² (0.83 kN/m²)
16.3 lb/ft² (0.79 kN/m²)
15.5 lb/ft² (0.75 kN/m²)
14.6 lb/ft² (0.70 kN/m²)
13.4 lb/ft² (0.65 kN/m²)
12.0 lb/ft² (0.58 kN/m²)
11.1 lb/ft² (0.53 kN/m²)

10.7 lb/ft² (0.52 kN/m²)

E-W wind

FIGURE 3.29 Horizontal wind pressures on the MWFRS for the building in Example 3.6.

Wind Direction	e_Q	$I_{\bar{z}}$	g_Q	Q	g_R	R	e
N-S	$0.15 \times 91.83 = 13.8$ ft (4.2 m)	0.25	3.4	0.84	4.0	0.51	12.3 ft (3.7 m)
E-W	$0.15 \times 61 = 9.2$ ft (2.8 m)	0.25	3.4	0.85	4.0	0.58	8.0 ft (2.4 m)

TABLE 3.29   Calculations for Eccentricity, e, for the Building in Example 3.6

The wind pressure and torsional moment at the mean roof height of the building for case 2 are as follows:

- N-S wind

$$0.75p_{Wy} = 0.75 \times 20.7 = 15.5 \text{ lb/ft}^2$$

$$0.75p_{Ly} = 0.75 \times 13.0 = 9.8 \text{ lb/ft}^2$$

$$M_T = 0.75(p_{Wy} + p_{Ly})B_y e_y$$

$$= 0.75 \times (20.7 + 13.0) \times 91.83 \times 12.3/1,000 = 28.6 \text{ ft-kips/ft}$$

In S.I.:

$$0.75p_{Wy} = 0.75 \times 1.00 = 0.75 \text{ kN/m}^2$$

$$0.75p_{Ly} = 0.75 \times 0.63 = 0.47 \text{ kN/m}^2$$

$$M_T = 0.75(p_{Wy} + p_{Ly})B_y e_y$$

$$= 0.75 \times (1.00 + 0.63) \times 28.0 \times 3.7 = 126.7 \text{ kN-m/m}$$

- E-W wind

$$0.75p_{Wx} = 0.75 \times 21.5 = 16.1 \text{ lb/ft}^2$$

$$0.75p_{Lx} = 0.75 \times 10.7 = 8.0 \text{ lb/ft}^2$$

$$M_T = 0.75(p_{Wx} + p_{Lx})B_x e_x$$

$$= 0.75 \times (21.5 + 10.7) \times 61.0 \times 8.0 / 1,000 = 11.8 \text{ ft-kips/ft}$$

In S.I.:

$$0.75p_{Wx} = 0.75 \times 1.04 = 0.78 \text{ kN/m}^2$$

$$0.75p_{Lx} = 0.75 \times 0.52 = 0.39 \text{ kN/m}^2$$

$$M_T = 0.75(p_{Wx} + p_{Lx})B_x e_x$$

$$= 0.75 \times (1.04 + 0.52) \times 18.6 \times 2.4 = 52.2 \text{ kN-m/m}$$

In case 3, 75 percent of the windward and leeward wall pressures in Fig. 3.29 act simultaneously on the building at each level above ground.

In case 4, 75 percent of the wind pressures and torsional moments defined in case 2 act simultaneously on the building.

Wind pressures and torsional moments for load cases 1 through 4 at the mean roof height of the building are given in Fig. 3.30.

The minimum design wind loading prescribed in ASCE/SEI 27.1.5 must be considered as a load case in addition to the load cases above.

3.4.7   Example 3.7—Wind Pressures on an Open Agricultural Building, MWFRS, Chapter 27, Part 1

Determine the wind pressures in both directions on the MWFRS of the open agricultural building in Fig. 3.31 using the requirements in Part 1 of Chapter 27 and the design data in Table 3.30. Assume obstructed wind flow beneath the roof and no fascia panels.

Solution

Check if the building meets all the conditions in ASCE/SEI 27.1.2, so that Part 1 in Chapter 27 can be used to determine the design wind pressures on the MWFRS.

The building is regular-shaped and does not have any unusual geometric irregularities in spatial form. Also, the building does not have any response characteristics that make it subject to across-wind loading or similar effects, and it is not sited at a location where channeling effects or buffeting in the wake of upwind obstructions need to be considered.

Therefore, the conditions in ASCE/SEI 27.1.2 are met and the requirements in Part 1 of Chapter 27 may be used.

The design procedure in Fig. 3.4 is used to determine the design wind pressures, p, in both directions.

> *Step 1—Determine the velocity pressure, q_h, at the mean roof height*   *Fig. 2.3*
>
>   The flowchart in Fig. 2.3 is used to determine q_z.
>
> - Step 1a—Determine the surface roughness category   ASCE/SEI 26.7.2
>
>   In Table 3.30, the surface roughness is given as C.
>
> - Step 1b—Determine the exposure category   ASCE/SEI 26.7.3
>
>   It is assumed that surface roughness C applies in all directions and that exposures B and D are not applicable. Therefore, the exposure category is C.
>
> - Step 1c—Determine the terrain exposure constants ASCE/SEI Table 26.11-1
>
>   For Exposure C, $\alpha = 9.5$ and $z_g = 900$ ft (274.32 m).
>
> - Step 1d—Determine the velocity pressure exposure coefficient, K_h
>
>   ASCE/SEI Table 26.10-1

$$\text{Mean roof height} = \frac{36 + 20}{2} = 28 \text{ ft (8.5 m)}$$

$$K_h = 2.01(z/z_g)^{2/\alpha} = 2.01 \times (28/900)^{2/9.5} = 0.97$$

In S.I.:

$$K_h = 2.01 \times (8.5/274.32)^{2/9.5} = 0.97$$

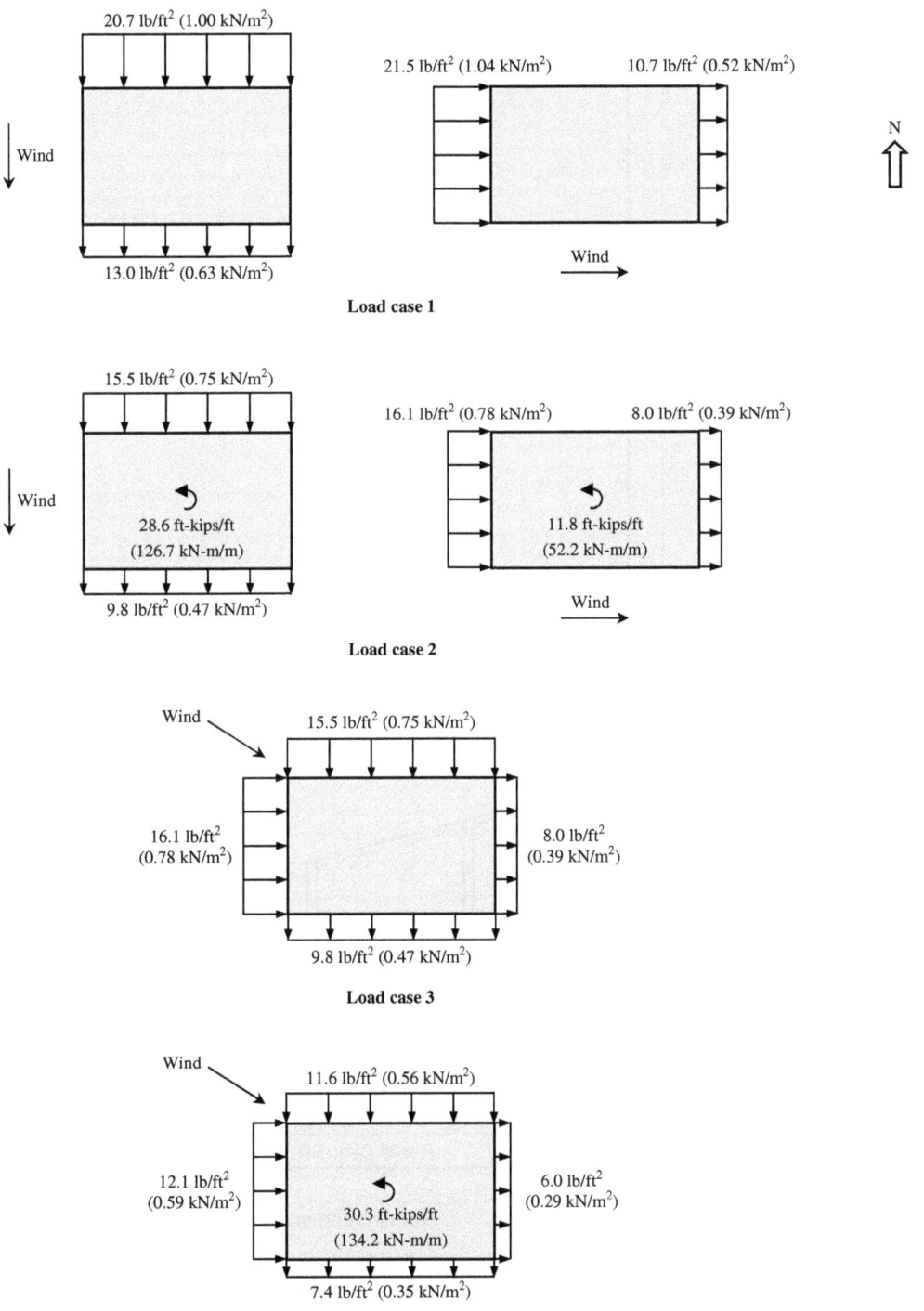

FIGURE 3.30 Load cases 1 through 4 at the mean roof height of the building in Example 3.6.

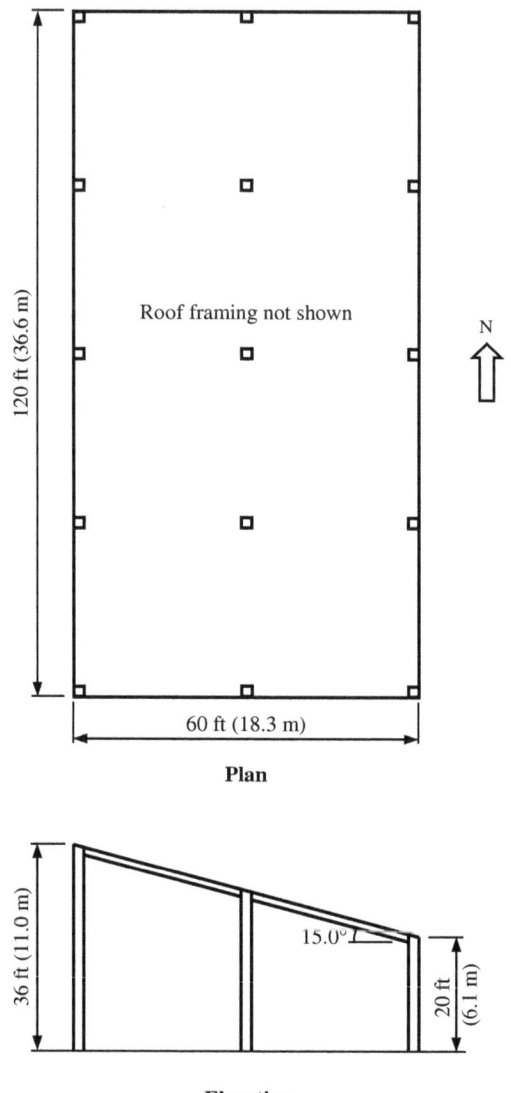

FIGURE 3.31 Plan and elevation of the open agricultural building in Example 3.7.

Location	Rapid City, SD
Surface roughness	C
Topography	Not situated on a hill, ridge, or escarpment
Occupancy	Low risk to human life in the event of failure
Enclosure classification	Open (no walls)
Fundamental natural frequency of the MWFRS	>1 Hz

TABLE 3.30 Design Data for the Open Agricultural Building in Example 3.7

- Step 1e—Determine the topographic factor, K_{zt} ASCE/SEI 26.8

  Because the building is not located on a hill, ridge, or escarpment, $K_{zt} = 1.0$.

- Step 1f—Determine the wind directionality factor, K_d ASCE/SEI Table 26.6-1

  For the MWFRS of a building structure, $K_d = 0.85$.

- Step 1g—Determine the ground elevation factor, K_e ASCE/SEI 26.9

  Ground elevation factor can be taken as 1.0 for all elevations.

- Step 1h—Determine the risk category of the building   ASCE/SEI Table 1.5-1

  Due to the nature of its occupancy, this agricultural building falls under Risk Category I.

- Step 1i—Determine the basic wind speed, V Table 2.1

  For Risk Category I, use IBC Figure 1609.3(4) or ASCE/SEI Figure 26.5-1A.

  Equivalently, use Ref. 3 or Ref. 4 to obtain $V = 105$ mi/h (47 m/s) for Rapid City, SD.

- Step 1j—Determine the wind velocity pressure, q_h ASCE/SEI 26.10.2

$$q_h = 0.00256 K_h K_{zt} K_d K_e V^2$$

$$= 0.00256 \times 0.97 \times 1.0 \times 0.85 \times 1.0 \times 105^2 = 23.3 \ \text{lb/ft}^2$$

In S.I.:

$$q_h = 0.613 K_h K_{zt} K_d K_e V^2$$

$$= 0.613 \times 0.97 \times 1.0 \times 0.85 \times 1.0 \times 47^2 / 1,000 = 1.12 \ \text{kN/m}^2$$

Step 2—Determine the gust-effect factor   *Fig. 2.4*

In the design data, $n_1 > 1$ Hz, which means the building is rigid.

It is permitted to take $G = 0.85$ for rigid buildings.

Step 3—Determine the net pressure coefficients, C_N   *ASCE/SEI Figures 27.3-4 and 27.3-7*

- Wind in the N-S direction ($\gamma = 90$ deg, 270 deg)

  ASCE/SEI Figure 27.3-7 is used to determine the net pressure coefficients, C_N, at various distances from the windward edge of the roof for obstructed wind flow (see Table 3.31).

- Wind in the E-W direction ($\gamma = 0$ deg, 180 deg)

  ASCE/SEI Figure 27.3-4 is used to determine the net pressure coefficients, C_{NW} and C_{NL}, on the windward and leeward portions of the roof surface for obstructed wind flow and a roof angle $\theta = 15$ deg (see Table 3.32).

Step 4—Determine the net design wind pressure p   *ASCE/SEI Equation (27.3-2)*

$$p = q_h G C_N = 23.3 \times 0.85 \times C_N = 19.8 C_N \ \text{lb/ft}^2$$

In S.I.:

$$p = 1.12 \times 0.85 \times C_N = 0.95 C_N \ \text{kN/m}^2$$

Horizontal Distance from Windward Edge	Load Case	C_N
$\leq h = 28$ ft (8.5 m)	A	−1.2
	B	0.5
$> h = 28$ ft (8.5 m), $\leq 2h = 56$ ft (17.1 m)	A	−0.9
	B	0.5
$> 2h = 56$ ft (17.1 m)	A	−0.6
	B	0.3

TABLE 3.31 Net Pressure Coefficients, C_N, for Wind in the N-S Direction

	West to East Wind ($\gamma = 0$ deg)		East to West Wind ($\gamma = 180$ deg)	
Load Case	C_{NW}	C_{NL}	C_{NW}	C_{NL}
A	−1.1	−1.5	0.4	−1.1
B	−2.1	−0.6	1.2	−0.3

TABLE 3.32 Net Pressure Coefficients, C_{NW} and C_{NL}, for Wind in the E-W Direction

Horizontal Distance from Windward Edge	Load Case	C_N	p, lb/ft² (kN/m²)
$\leq h = 28$ ft (8.5 m)	A	−1.2	−23.8 (−1.14)
	B	0.5	9.9 (0.48)
$> h = 28$ ft (8.5 m), $\leq 2h = 56$ ft (17.1 m)	A	−0.9	−17.8 (−0.86)
	B	0.5	9.9 (0.48)
$> 2h = 56$ ft (17.1 m)	A	−0.6	11.9 (−0.57)
	B	0.3	5.9 (0.29)

TABLE 3.33 Net Design Wind Pressures, p, for Wind in the N-S Direction

	West to East Wind ($\gamma = 0$ deg)				East to West Wind ($\gamma = 180$ deg)			
Load Case	C_{NW}	p, lb/ft² (kN/m²)	C_{NL}	p, lb/ft² (kN/m²)	C_{NW}	p, lb/ft² (kN/m²)	C_{NL}	p, lb/ft² (kN/m²)
A	−1.1	−21.8 (−1.05)	−1.5	−29.7 (−1.43)	0.4	7.9 (0.38)	−1.1	−21.8 (−1.05)
B	−2.1	−41.6 (−2.00)	−0.6	−11.9 (−0.57)	1.2	23.8 (1.14)	−0.3	−5.9 (−0.29)

TABLE 3.34 Net Design Wind Pressures, p, for Wind in the in the E-W Direction

Net design wind pressures for wind in the N-S and E-W directions are given in Tables 3.33 and 3.34, respectively. The pressures act perpendicular to the roof surface.

The minimum wind pressures in ASCE/SEI 27.1.5 must also be considered in addition to load cases A and B above (see Fig. 3.1).

Step 5—Determine additional wind pressures in accordance with ASCE/SEI 27.3.2

Because there are no fascia panels and the roof is not pitched, no additional wind pressures in accordance with ASCE/SEI 27.3.2 need to be determined.

CHAPTER 4

Wind Loads on Buildings: MWFRS (Envelope Procedure)

4.1 Overview

This chapter contains the requirements for determining wind pressures and loads on the main wind force resisting systems (MWFRSs) of low-rise buildings in accordance with the Envelope Procedure of ASCE/SEI Chapter 28. According to ASCE/SEI 26.2, a low-rise building is an enclosed or partially enclosed building that complies with the following conditions:

1. Mean roof height, h, is less than or equal to 60 ft (18.3 m).

2. Mean roof height, h, does not exceed the least horizontal dimension of the building.

A summary of the wind load procedures in Chapter 28 is given in Table 4.1.

Part 1 in Chapter 28 is applicable to enclosed, partially enclosed, and open low-rise buildings with flat, gable, or hip roofs. Wind pressures are determined based on wind direction using equations appropriate for each surface of the building. Part 2 is applicable to enclosed, simple diaphragm low-rise buildings (that is, buildings in which both windward and leeward wind loads are transmitted by roof and vertically spanning wall assemblies, through continuous floor and roof diaphragms, to the MWFRS) where wind pressures are determined directly from a table and are applied on horizontal and vertical projected surfaces of the building.

The provisions in Chapter 28 are applicable to buildings that comply with the following (ASCE/SEI 28.1.2):

- The building is regular-shaped, that is, the building has no unusual geometrical irregularities in spatial form.

- The building does not have response characteristics that make it subject to across-wind loading, vortex shedding, or instability caused by galloping or flutter. Additionally, the building is not located at a site where channeling effects or buffeting in the wake of upwind obstructions warrant special consideration.

Buildings not meeting these conditions must be designed by either recognized literature documenting such wind load effects or by the wind tunnel procedure in ASCE/SEI Chapter 31 (ASCE/SEI 28.1.3).

Reduction in wind pressure due to apparent shielding by surrounding buildings, other structures, or terrain features is not permitted (ASCE/SEI 28.1.4). Such shielding may be modified or completely removed during the lifespan of the building, which could result in significantly higher wind loads.

117

| | Applicability | | |
Part	Low-Rise Building Type	Height Limit	Conditions
1	Enclosed	$h \leq 60$ ft (18.3 m)	• Regular-shaped building • Building does not have response characteristics making it subject to across-wind loading, vortex shedding, instability due to galloping or flutter • Building is not located at a site where channeling effects or buffeting in the wake of upwind obstructions warrant special consideration
	Partially enclosed	$h \leq$ least horizontal dimension of building	
	Open		
2	Enclosed, simple diaphragm		• Same conditions as in Part 1 • $n_1 \geq 1$ Hz • Building has an approximately symmetrical cross section in each direction with either a flat roof or a gable or hip roof with $\theta \leq 45$ degrees • Building is exempted from torsional load cases as indicated in note 5 of ASCE/SEI Figure 28.3-1, or the torsional load cases defined in note 5 do not control the design of any of the MWFRSs of the building

TABLE 4.1 Wind Load Procedures in ASCE/SEI Chapter 28

4.2 Enclosed and Partially Enclosed Low-Rise Buildings (Part 1)

4.2.1 Design Wind Pressures, *p*, for Low-Rise Buildings

Design wind pressures, *p*, for the MWFRS of low-rise buildings satisfying the conditions of Part 1 of Chapter 28 are determined by ASCE/SEI Equation (28.3-1):

$$p = q_h[(GC_{pf}) - (GC_{pi})] \qquad (4.1)$$

This equation is used to calculate wind pressures on the building surfaces identified in ASCE/SEI Figure 28.3-1. The velocity pressure at the mean roof height of the building, q_h, is determined in accordance with ASCE/SEI 26.10 (see Sec. 2.7 of this publication). External pressure coefficients, (GC_{pf}), and internal pressure coefficients, (GC_{pi}), are determined from ASCE/SEI Figure 28.3-1 and ASCE/SEI Table 26.13-1, respectively. The combined gust-effect factor and external pressure coefficients, (GC_{pf}), which were determined from wind tunnel tests, are not permitted to be separated (ASCE/SEI 28.3.1.1).

A building must be designed for all wind directions by considering in turn each corner of the building as the windward corner (see ASCE/SEI Figure 28.3-1). At each corner, two load cases must be considered (load case A and load case B), one for each range of wind direction. In general, a total of 16 separate load cases must be evaluated because both positive and negative internal pressures must be considered. For symmetrical buildings, some of these load cases are repetitive and can be

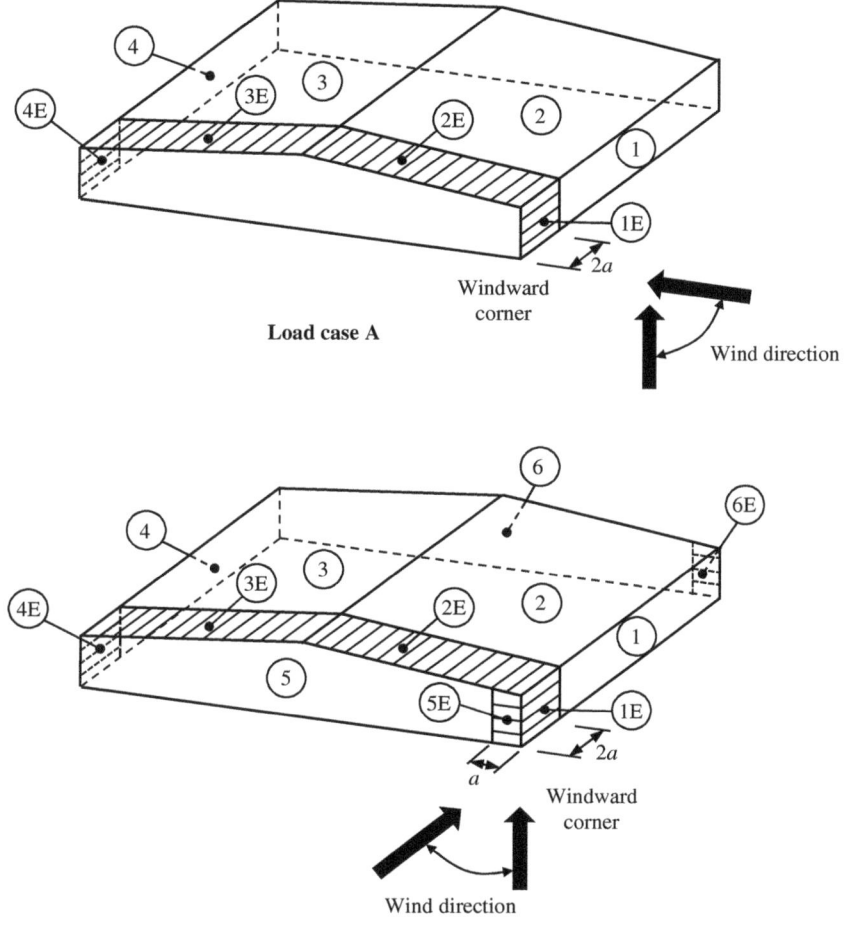

Load case A

Windward corner

Wind direction

Load case B

Windward corner

Wind direction

FIGURE 4.1   Basic load cases for low-rise buildings designed in accordance with Part 1 of ASCE/SEI Chapter 28.

eliminated. Illustrated in Fig 4.1 are load cases A and B for the same windward corner of a low-rise building.

A suggested method to determine wind pressures using ASCE/SEI Figure 28.3-1 for low-rise buildings with hip roofs is given in ASCE/SEI C28.3-2.

The torsional load cases in ASCE/SEI Figure 28.3-1 must be considered in the design of all low-rise buildings except for the following (see note 5 in this figure):

- One-story buildings with a mean roof height of less than or equal to 30 ft (9.1 m).
- Buildings two stories or less framed with light-frame construction.
- Buildings two stories or less with flexible diaphragms.

A step-by-step procedure to determine the design wind pressures on enclosed, partially enclosed, and open low-rise buildings with flat, gable, or hip roofs in accordance with Part 1 of ASCE/SEI Chapter 28 is given in Fig. 4.2.

4.2.2 Parapets

Design wind pressures for the effects of parapets, p_p, on the MWFRS of low-rise buildings with flat, gable, or hip roofs are determined by ASCE/SEI Equation (28.3-2):

$$p_p = q_p(GC_{pn}) \tag{4.2}$$

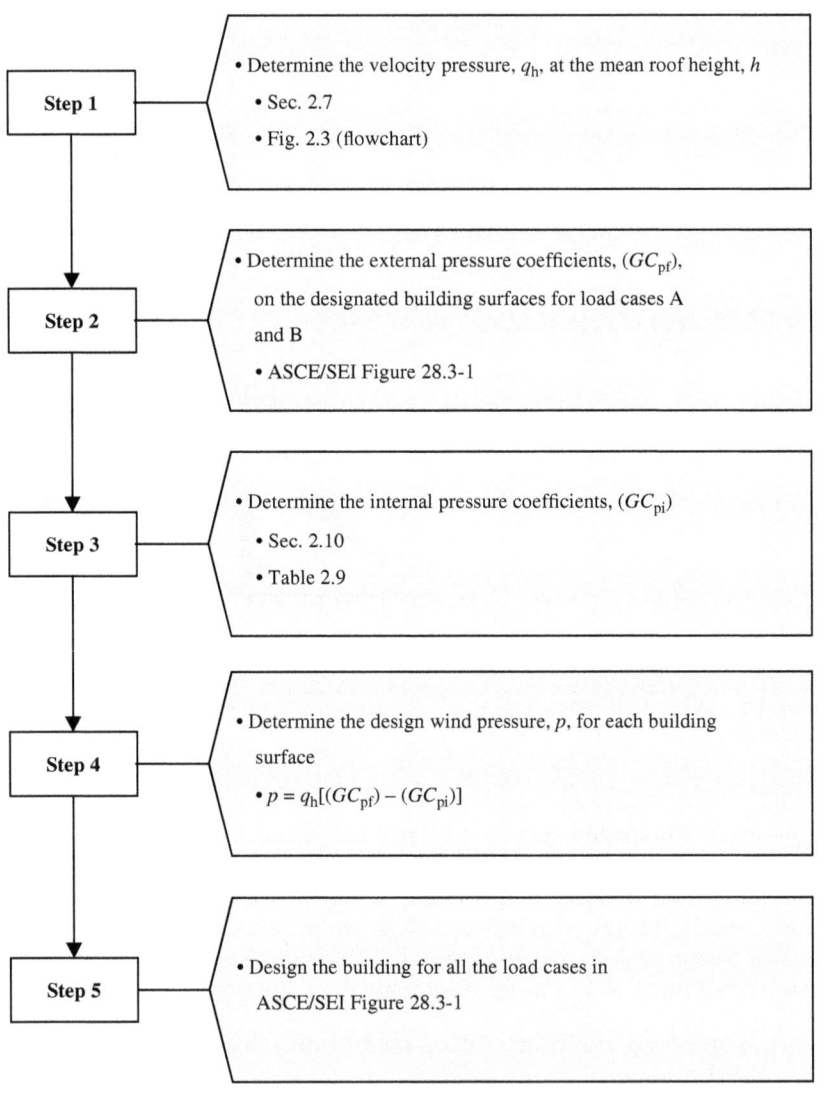

FIGURE 4.2 Procedure to determine design wind pressures, p, on enclosed, partially enclosed, and open low-rise buildings with flat, gable, or hip roofs (Envelope Procedure, Part 1).

In this equation, q_p is the velocity pressure evaluated at the top of the parapet determined in accordance with ASCE/SEI 26.10 (see Sec. 2.7 of this publication) and (GC_{pn}) is the combined net pressure coefficient, which is equal to +1.5 for a windward parapet and −1.0 for a leeward parapet (see Fig. 3.7 of this publication).

4.2.3 Roof Overhangs

Positive external pressures on the bottom surface of windward roof overhangs are determined using the pressure coefficient $GC_p = 0.7$ in combination with the top surface pressures determined by ASCE/SEI Figure 28.3-1 (ASCE/SEI 28.3.3). Application of this pressure is similar to that shown in Fig. 3.5 of this publication.

Provisions are not provided for wind pressures on the bottom surface of a leeward overhang.

4.2.4 Minimum Design Wind Loads

Minimum design wind pressures in the design of the MWFRS for enclosed or partially enclosed low-rise buildings are given in ASCE/SEI 28.3.4. The pressures of 16 lb/ft² (0.77 kN/m²) on the projected area of the walls and 8 lb/ft² (0.38 kN/m²) on the projected area of the roof are considered a separate load case from any of the other load cases required in Part 1 (see Fig. 3.1).

4.2.5 Horizontal Wind Loads on Open or Partially Enclosed Buildings with Transverse Frames and Pitched Roofs

The horizontal pressure, p, in the direction parallel to the roof ridge of an open or partially enclosed building with transverse frames and a pitched roof with an angle less than or equal to 45 degrees is determined by ASCE/SEI Equation (28.3-3):

$$p = q_h[(GC_{pf})_{windward} - (GC_{pf})_{leeward}]K_B K_S \tag{4.3}$$

This equation is applicable to buildings with open end walls and with end walls fully or partially enclosed with cladding. The pressures determined by Eq. (4.3) act in combination with the pressures on the roof calculated in accordance with ASCE/SEI 27.3.2 for an open or partially enclosed building (see Sec. 3.2.2 of this publication).

On the windward end wall, the external pressure coefficients $(GC_{pf})_{windward}$ are obtained from load case B in ASCE/SEI Figure 28.3-1 for surfaces 5 and 5E. Similarly, for the leeward wall, external pressure coefficients $(GC_{pf})_{leeward}$ are obtained from load case B in ASCE/SEI Figure 28.3-1 for surfaces 6 and 6E. The terms K_B and K_S are the frame width factor and shielding factor, respectively, and are determined by the equations in ASCE/SEI 28.3.5.

The frame width factor, K_B, is determined as follows:

$$K_B = \begin{cases} 1.8\text{-}0.01B \text{ where } B < 100 \text{ ft (30.5 m)} \\ \\ 0.8 \text{ where } B \geq 100 \text{ ft (30.5 m)} \end{cases} \tag{4.4}$$

where B is the width of the building perpendicular to the ridge in feet (meters).

The frame shielding factor, K_S, is determined based on the number of frames, n, perpendicular to the direction of analysis, and the solidity ratio, ϕ, which is the ratio of the effective solid area of an end wall, A_S (that is, the projected area of any portion of the end wall exposed to the wind), to the total end wall area for an equivalent closed building, A_E (see ASCE/SEI Figure 28.3-2):

$$K_S = 0.60 + 0.073(n - 3) + 1.25\phi^{1.8} \tag{4.5}$$

In buildings with two frames, n must be taken as 3 (ASCE/SEI 28.3.5).

The total longitudinal force, F, that must be resisted by the MWFRS due to p is determined by ASCE/SEI Equation (28.3-4):

$$F = pA_E \tag{4.6}$$

This force is applied at the centroid of A_E.

The flowchart in Fig. 4.3 can be used to determine p and F.

In the direction normal to the ridge, the wind pressures are calculated in accordance with ASCE/SEI 27.3.2 (see Sec. 3.2.2 of this publication). These pressures are a separate load case from the pressures determined parallel to the ridge, as discussed above.

4.3   Enclosed Simple Diaphragm Low-Rise Buildings (Part 2)

4.3.1   Overview

A method for determining wind pressures for the MWFRS of enclosed, simple diaphragm low-rise buildings with flat, gable, or hip roofs meeting the conditions in ASCE/SEI 28.5.2 is given in Part 2 of Chapter 28 (see Table 4.1 of this publication).

A simple diaphragm building is one in which both windward and leeward wind loads are transmitted by roof and vertically spanning wall assemblies through continuous floor and roof diaphragms to the MWFRS (ASCE/SEI 26.2). As such, internal wind pressures cancel out in the determination of the total wind load in the direction of analysis. Buildings with structural expansion joints in diaphragms or structural systems with girts or other horizontal members that transfer significant wind loads directly to vertical members of the MWFRS are not permitted to be designed using the methods in Part 2.

4.3.2   Design Wind Pressures

Design wind pressures, p_{s30}, for walls (zones A and C), roofs (zones B, D, and E through H), and overhangs are tabulated in ASCE/SEI Figure 28.5-1 as a function of the basic wind speed, V, and roof angle, θ, for buildings with a mean roof height, h, of 30 ft (9.1 m) located on primarily flat ground in Exposure B. Modifications are made to the tabulated pressures based on actual building height and exposure using the adjustment factor, λ, given in the figure. Such pressures must also be modified by the topographic factor, K_{zt}, in accordance with ASCE/SEI 26.8 where applicable.

Wind pressure, p_s, is determined by ASCE/SEI Equation (28.5-1):

$$p_s = \lambda K_{zt} p_{s30} \tag{4.7}$$

Horizontal wall pressures on zones A and C are the net sum of the windward and leeward pressures on vertical projections of the wall, as shown in ASCE/SEI Figure 28.5-1 (see Fig. 4.4). Horizontal roof pressures on zones B and D are the net sum of the

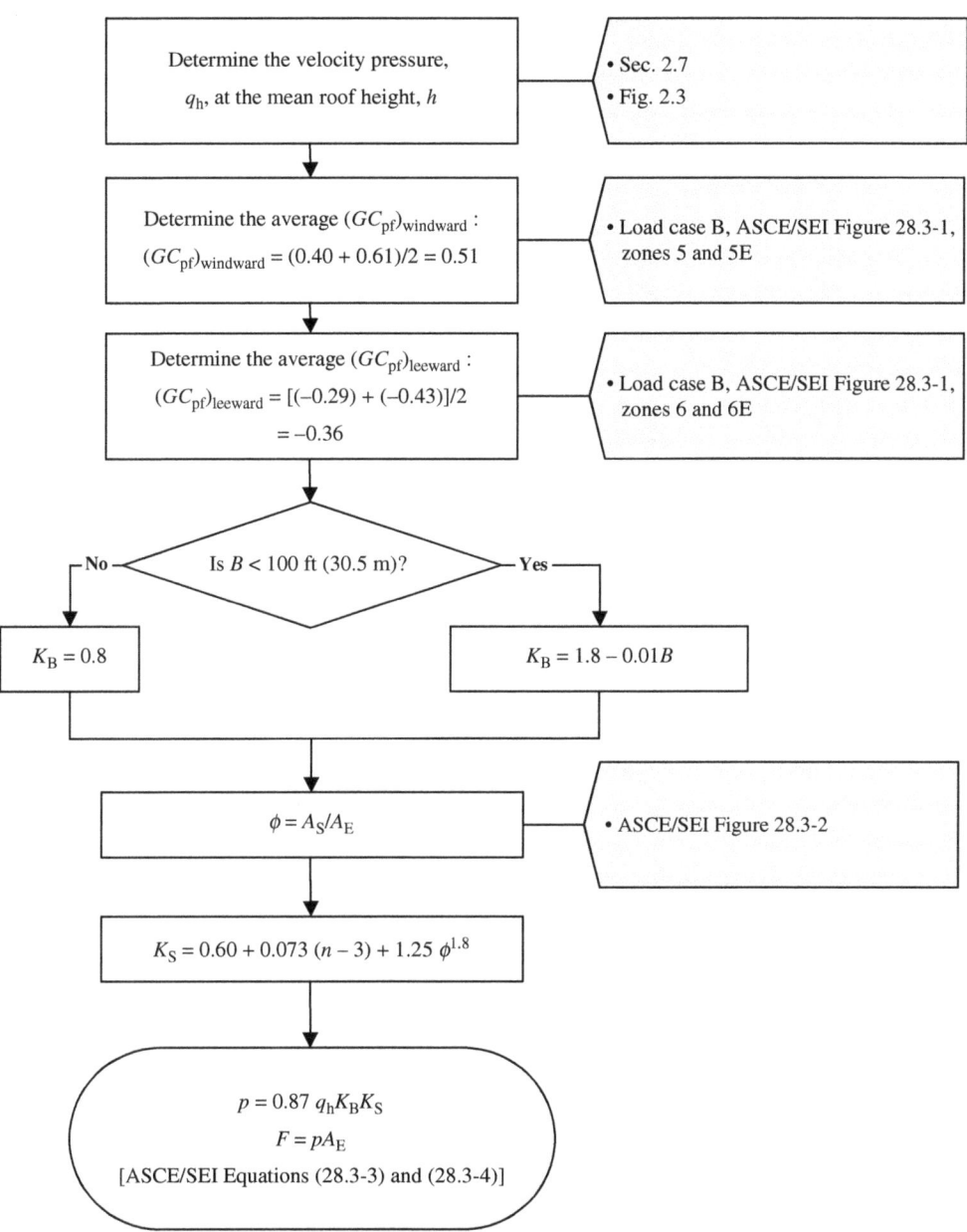

FIGURE 4.3 Flowchart to determine horizontal wind loads on open and partially enclosed buildings with transverse frames and pitched roofs (Envelope Procedure, Part 1).

windward and leeward pressures on the vertical projection of the roof, and the vertical roof pressures on zones E through H are the net sum of the external and internal pressures (using an internal pressure coefficient of ±0.18 for enclosed buildings) on the horizontal projection of the roof.

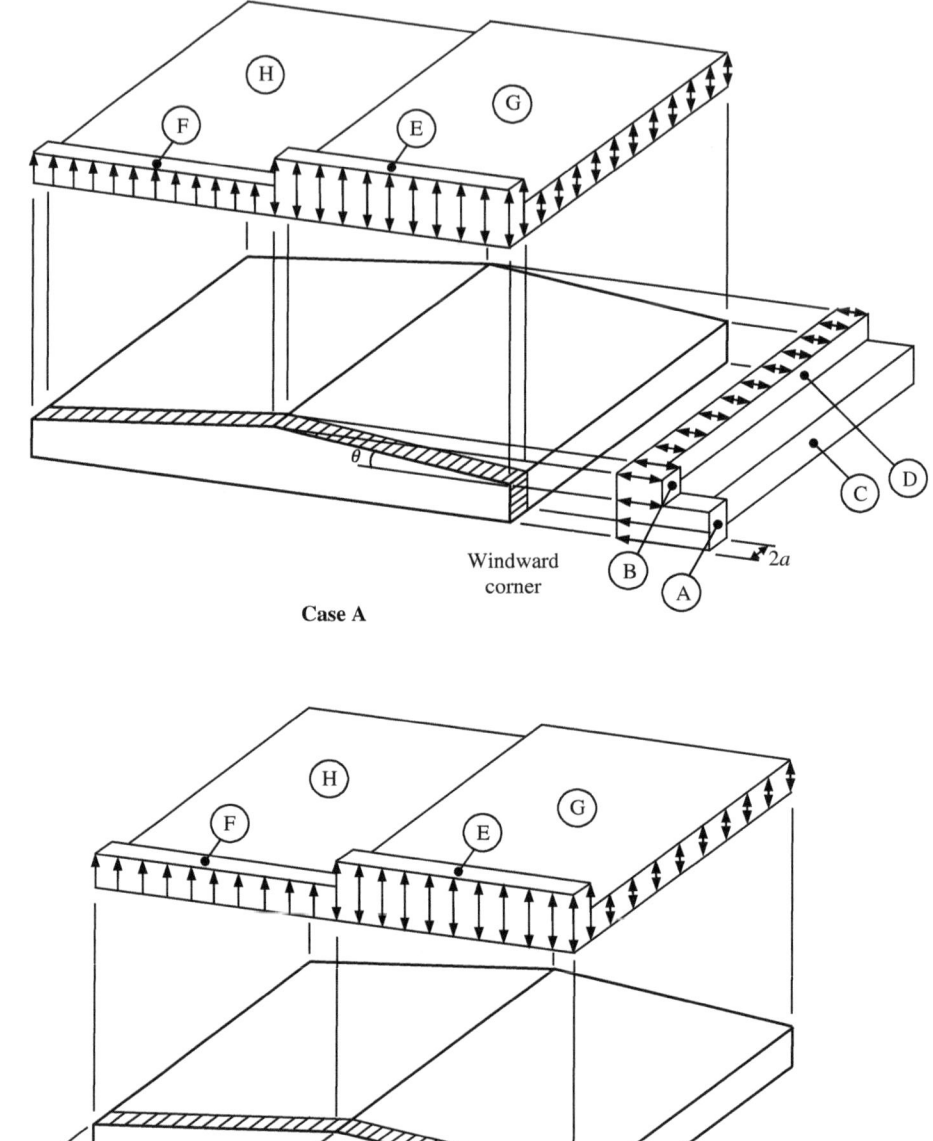

Case A

Case B

FIGURE 4.4   Wind pressures in accordance with Part 2 of the Envelope Procedure.

The total horizontal load must not be decreased due to negative pressures in zones B and D (see Note 7 in ASCE/SEI Figure 28.5-1). Where these pressures are negative, p_s must be taken as zero in these zones.

Design wind pressures on windward and leeward walls are obtained by multiplying p_s in zones A and C by +0.85 and −0.70, respectively. For sidewalls, the pressure is determined by multiplying p_s in zone C by −0.65.

The load patterns for each load case shown in ASCE/SEI Figure 28.5-1 must be applied to each corner of a building in turn as the windward corner, similar to ASCE/SEI Figure 28.3-1.

4.3.3 Minimum Design Wind Loads

Minimum design wind loads are prescribed in ASCE/SEI 28.5.4 as follows assuming $p_s = 0$ in zones E through H:

- Zones A and C: +16 lb/ft² (0.77 kN/m²)
- Zones B and D: +8 lb/ft² (0.38 kN/m²)

The step-by-step procedure in Fig. 4.5 can be used to determine the design pressures using Part 2 of Chapter 28.

4.4 Examples

The following examples illustrate the determination of wind pressures using Parts 1 and 2 of the Envelope Procedure.

4.4.1 Example 4.1—Wind Pressures on a Commercial Building, MWFRS, Chapter 28, Part 1

Determine the wind pressures in both directions on the MWFRS of the commercial building in Fig. 3.13 using the requirements in Part 1 of Chapter 28 and the design data in Table 3.2.

Solution

Check if the building meets all the conditions in ASCE/SEI 28.1.2 so that Part 1 in Chapter 28 can be used to determine the design wind pressures on the MWFRS.

The building is regular-shaped and does not have any unusual geometric irregularities in spatial form. Also, the building does not have any response characteristics that make it subject to across-wind loading or similar effects, and it is not sited at a location where channeling effects or buffeting in the wake of upwind obstructions need to be considered.

The following two conditions must be checked to determine if the building is a low-rise building (ASCE/SEI 26.2):

- Mean roof height = 36 ft (11.0 m) < 60 ft (18.3 m)
- Mean roof height = 36 ft (11.0 m) < least horizontal dimension = 50 ft (15.2 m)

Thus, the building is a low-rise building.

Therefore, the conditions in ASCE/SEI 28.1.2 are met and the requirements in Part 1 of Chapter 28 may be used.

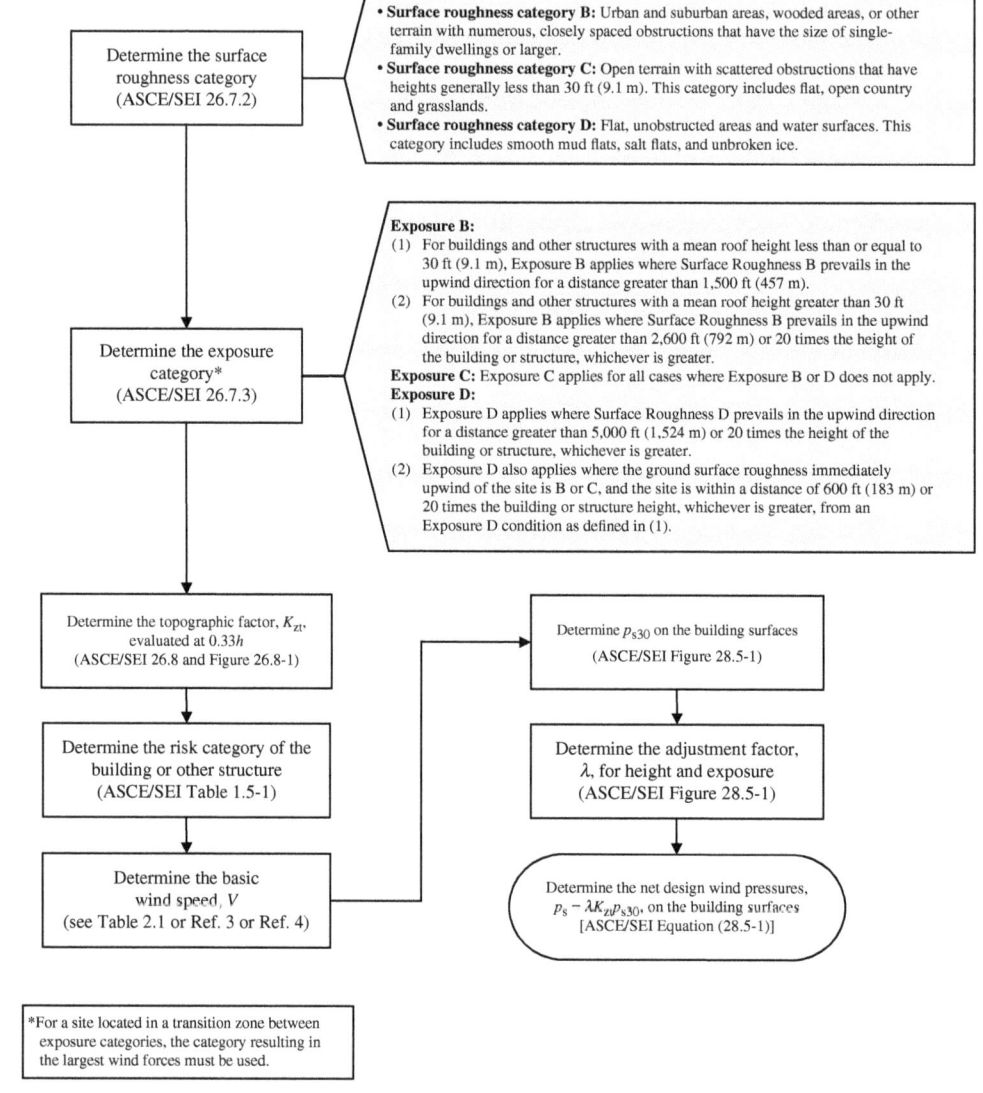

FIGURE 4.5 Procedure to determine design wind pressures on enclosed simple diaphragm low-rise buildings (Envelope Procedure, Part 2).

The design procedure in Fig. 4.2 is used to determine the design wind pressures, p, in both directions.

    Step 1—Determine the velocity pressure, q_h, at the mean roof height   Fig. 2.3

        The flowchart in Fig. 2.3 is used to determine q_h.

        • Step 1a—Determine the surface roughness category   ASCE/SEI 26.7.2

        In Table 3.2, the surface roughness is given as C.

- Step 1b—Determine the exposure category   ASCE/SEI 26.7.3

It is assumed that surface roughness C applies in all directions and that exposures B and D are not applicable. Therefore, the exposure category is C.

- Step 1c—Determine the terrain exposure constants   ASCE/SEI Table 26.11-1

For Exposure C, $\alpha = 9.5$ and $z_g = 900$ ft (274.32 m).

- Step 1d—Determine the velocity pressure exposure coefficient, K_h
                                                ASCE/SEI Table 26.10-1

At $z = 36$ ft (11.0 m):

$$K_h = 2.01(z/z_g)^{2/\alpha} = 2.01 \times (36/900)^{2/9.5} = 1.02$$

In S.I.:

$$K_h = 2.01 \times (11.0/274.32)^{2/9.5} = 1.02$$

- Step 1e—Determine the topographic factor, K_{zt} ASCE/SEI 26.8

Because the building is not located on a hill, ridge, or escarpment, $K_{zt} = 1.0$.

- Step 1f—Determine the wind directionality factor, K_d
                                                ASCE/SEI Table 26.6-1

For the MWFRS of a building structure, $K_d = 0.85$.

- Step 1g—Determine the ground elevation factor, K_e ASCE/SEI 26.9

Ground elevation factor can be taken as 1.0 for all elevations.

- Step 1h—Determine the risk category of the building
                                                ASCE/SEI Table 1.5-1

Due to the nature of its occupancy, this commercial building falls under Risk Category II.

- Step 1i—Determine the basic wind speed, V Table 2.1

For Risk Category II, use IBC Figure 1609.3(1) or ASCE/SEI Figure 26.5-1B. Equivalently, use Ref. 3 or Ref. 4 to obtain $V = 101$ mi/h (45 m/s) for Phoenix, AZ.

- Step 1j—Determine the wind velocity pressure, q_h ASCE/SEI 26.10.2

$$q_h = 0.00256 \, K_h K_{zt} K_d K_e V^2$$

$$= 0.00256 \times 1.02 \times 1.0 \times 0.85 \times 1.0 \times 101^2 = 22.6 \text{ lb/ft}^2$$

In S.I.:

$$q_h = 0.613 K_h K_{zt} K_d K_e V^2$$

$$= 0.613 \times 1.02 \times 1.0 \times 0.85 \times 1.0 \times 45^2/1{,}000 = 1.08 \text{ kN/m}^2$$

Surface	(GC_pf) Load Case A	Load Case B
1	0.40	−0.45
2	−0.69	−0.69
3	−0.37	−0.37
4	−0.29	−0.45
5	—	0.40
6	—	−0.29
1E	0.61	−0.48
2E	−1.07	−1.07
3E	−0.53	−0.53
4E	−0.43	−0.48
5E	—	0.61
6E	—	−0.43

TABLE 4.2   External Pressure Coefficients, (GC_{pf}), for the Building in Example 4.1

Step 2—Determine the external pressure coefficients, (GC_{pf}) ASCE/SEI Figure 28.3-1

External pressure coefficients, (GC_{pf}), are read directly from ASCE/SEI Figure 28.3-1 using a roof angle between 0 and 5 degrees for wind in the N-S direction (load case A) and roof angle between 0 and 90 degrees for wind in the E-W direction (load case B). A summary of the pressure coefficients for both load cases is given in Table 4.2.

Step 3—Determine the internal pressure coefficient, (GC_{pi}) Table 2.9

For an enclosed building, (GC_{pi}) = +0.18, −0.18.

Step 4—Determine the design wind pressures, p   ASCE/SEI Equation (28.3-1)

$$p = q_h[(GC_{pf}) - (GC_{pi})] = 22.6 \times [(GC_{pf}) - (\pm 0.18)]$$

In S.I.:

$$p = q_h[(GC_{pf}) - (GC_{pi})] = 1.08 \times [(GC_{pf}) - (\pm 0.18)]$$

Calculations for design wind pressure are illustrated for surface 1 for wind in the N-S and E-W directions:

- N-S direction (load case A)

  For positive internal pressure: $p = 22.6 \times (0.40 - 0.18) = 5.0$ lb/ft^2

  For negative internal pressure: $p = 22.6 \times [0.40 - (-0.18)] = 13.1$ lb/ft^2

- E-W direction (load case B)

  For positive internal pressure: $p = 22.6 \times [(-0.45) - 0.18)] = -14.2 \text{ lb/ft}^2$

  For negative internal pressure: $p = 22.6 \times [(-0.45) - (-0.18)] = -6.1 \text{ lb/ft}^2$

In S.I.:

- N-S direction (load case A)

  For positive internal pressure: $p = 1.08 \times (0.40 - 0.18) = 0.24 \text{ kN/m}^2$

  For negative internal pressure: $p = 1.08 \times [0.40 - (-0.18)] = 0.63 \text{ kN/m}^2$

- E-W direction (load case B)

  For positive internal pressure: $p = 1.08 \times [(-0.45) - 0.18)] = -0.68 \text{ kN/m}^2$

  For negative internal pressure: $p = 1.08 \times [(-0.45) - (-0.18)] = -0.29 \text{ kN/m}^2$

A summary of the design wind pressures is given in Table 4.3.

The distance a is equal to the following:

$$a = \text{smaller of} \begin{cases} 0.1 \times \text{least horizontal dimension} = 0.1 \times 50 = 5 \text{ ft (1.5 m)} \\ 0.4\, h = 0.4 \times 36 = 14.4 \text{ ft (4.4 m)} \end{cases}$$

	Design Wind Pressure, p, lb/ft² (kN/m²)					
	Load Case A (N-S Wind)			Load Case B (E-W Wind)		
		(GC_{pi})			(GC_{pi})	
Surface	(GC_{pf})	$+(GC_{pi})$	$-(GC_{pi})$	(GC_{pf})	$+(GC_{pi})$	$-(GC_{pi})$
1	0.40	5.0 (0.24)	13.1 (0.63)	-0.45	-14.2 (-0.68)	-6.1 (-0.29)
2	-0.69	-19.7 (-0.94)	-11.5 (-0.55)	-0.69	-19.7 (-0.94)	-11.5 (-0.55)
3	-0.37	-12.4 (-0.59)	-4.3 (-0.21)	-0.37	-12.4 (-0.59)	-4.3 (-0.21)
4	-0.29	-10.6 (-0.51)	-2.5 (-0.12)	-0.45	-14.2 (-0.68)	-6.1 (-0.29)
5	—	—	—	0.40	5.0 (0.24)	13.1 (0.63)
6	—	—	—	-0.29	-10.6 (-0.51)	-2.5 (-0.12)
1E	0.61	9.7 (0.46)	17.9 (0.85)	-0.48	-14.9 (-0.71)	-6.8 (-0.32)
2E	-1.07	-28.3 (-1.35)	-20.1 (-0.96)	-1.07	-28.3 (-1.35)	-20.1 (-0.96)
3E	-0.53	-16.1 (-0.77)	-7.9 (-0.38)	-0.53	-16.1 (-0.77)	-7.9 (-0.38)
4E	-0.43	-13.8 (-0.66)	-5.7 (-0.27)	-0.48	-14.9 (-0.71)	-6.8 (-0.32)
5E	—	—	—	0.61	9.7 (0.46)	17.9 (0.85)
6E	—	—	—	-0.43	-13.8 (-0.66)	-5.7 (-0.27)

TABLE 4.3 Design Wind Pressures on the MWFRS of the Building in Example 4.1

Minimum a = greater of

$$\begin{cases} 0.04 \times \text{least horizontal dimension} = 0.04 \times 50 = 2 \text{ ft (0.61 m)} \\ 3 \text{ ft (0.91 m)} \end{cases}$$

Therefore, $a = 5$ ft (1.5 m).

For flat roofs, the boundary for zone 2/3 and zone 2E/3E must be at the mid-width of the building (see note 7 in in ASCE/SEI Figure 28.3-1). Therefore, in the N-S direction, the boundary is located 25 ft (7.6 m) from the windward edge of the roof. In the E-W direction, the boundary is located 37.5 ft (11.4 m) from the windward edge of the roof.

The roof pressure coefficients, (GC_{pf}), are negative in zones 2 and 2E (see Table 4.2 or Table 4.3). According to note 8 in ASCE/SEI Figure 28.3-1, these pressure coefficients must be applied in zone 2/2E for a distance from the edge of the roof equal to 50 percent of the horizontal dimension of the building parallel to the direction of the MWFRS being designed or 2.5 times the eave height, h_e, at the windward wall, whichever is less. The remainder of the zone 2/2E extending to the ridge line must use the pressure coefficients for zone 3/3E.

For this building:

- N-S direction: $0.5 \times 50 = 25.0$ ft (7.6 m)
- E-W direction: $0.5 \times 75 = 37.5$ ft (11.4 m)

$$2.5 \, h_e = 2.5 \times 36.0 = 90.0 \text{ ft (27.4 m)}$$

Therefore, in the N-S direction, zone 2/2E applies over a distance of 25.0 ft (7.6 m) and zone 3/3E applies over a distance $50.0 - 25.0 = 25.0$ ft (7.6 m). In the E-W direction, zone 2/2E applies over a distance of 37.5 ft (11.4 m) from the edge of the windward roof, and zone 3/3E applies over a distance of $75.0 - 37.5 - 37.5$ ft (11.4 m).

Step 5—Design the building for all the applicable load cases   ASCE/SEI Figure. 28.3-1

According to note 4 in ASCE/SEI Figure 28.3-1, combinations of external and internal pressures must be evaluated to obtain the most severe loading. In general, the building must be designed for the eight load patterns in ASCE/SEI Figure 28.3-1. Considering both positive and negative internal pressures for each load pattern results in a total of 16 separate load cases. Because the building is symmetrical in the N-S direction and in the E-W direction, the 16 load cases reduces to 4.

Design wind pressures in the N-S and E-W directions for positive and negative internal pressures are given in Figs. 4.6 and 4.7, respectively (in both cases, the windward corner is taken at the southwest corner of the building).

The torsional load cases in ASCE/SEI Figure 28.3-1 are also applicable for this building because it does not conform with any of the conditions in the exception to note 5 in the figure. The pressures in zones designated with a "T" are equal to 25 percent of the full design wind pressures in zones 1 through 6. Torsional loading is applicable to all eight basic load patterns where each corner is considered the windward corner.

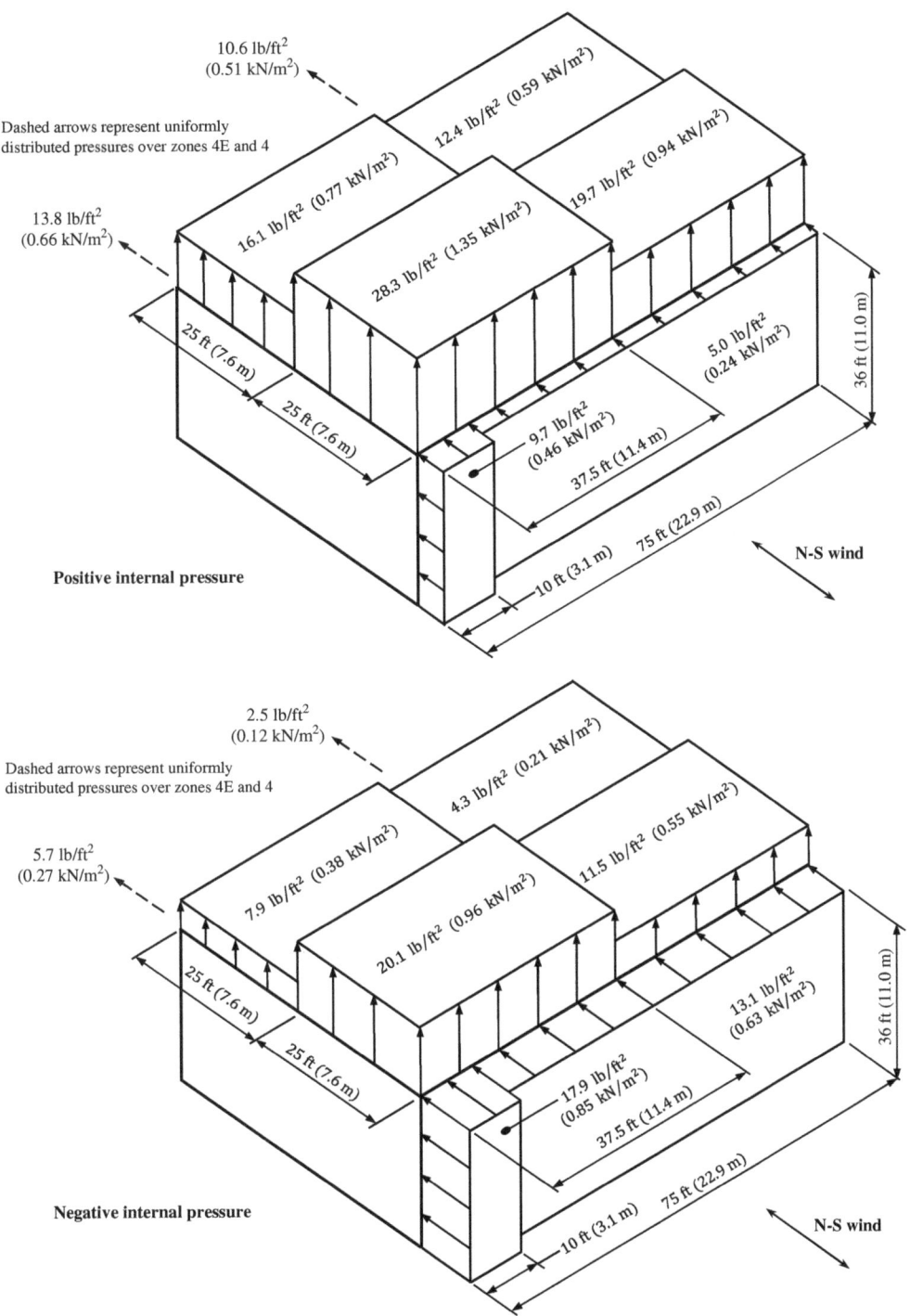

Figure 4.6 Design wind pressures in the N-S direction (load case A) for the building in Example 4.1.

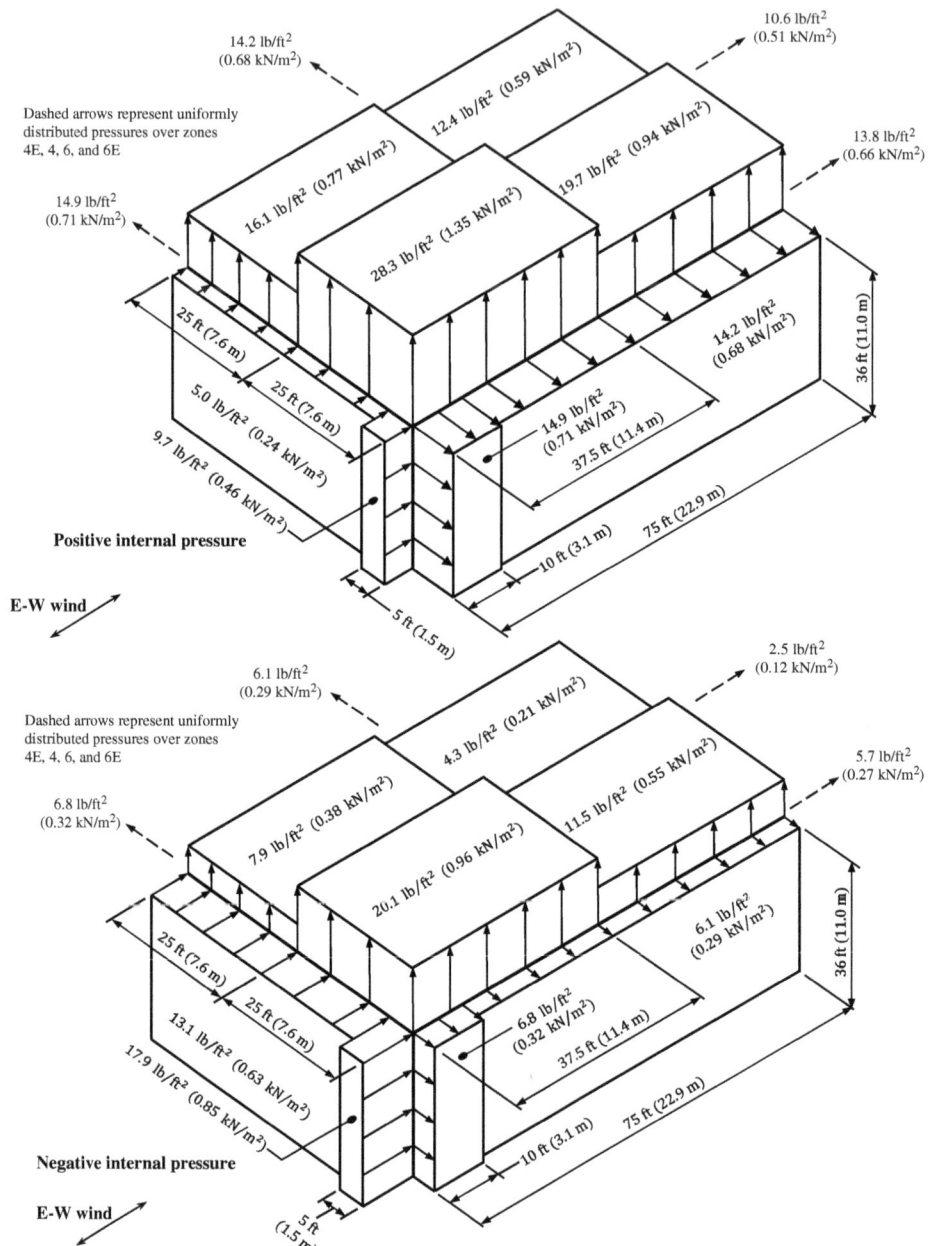

Figure 4.7 Design wind pressures in the E-W direction (load case B) for the building in Example 4.1.

Design wind pressures in the N-S and E-W directions considering torsional loading for positive and negative internal pressures are given in Figs. 4.8 and 4.9, respectively (in both cases, the windward corner is taken at the southwest corner of the building).

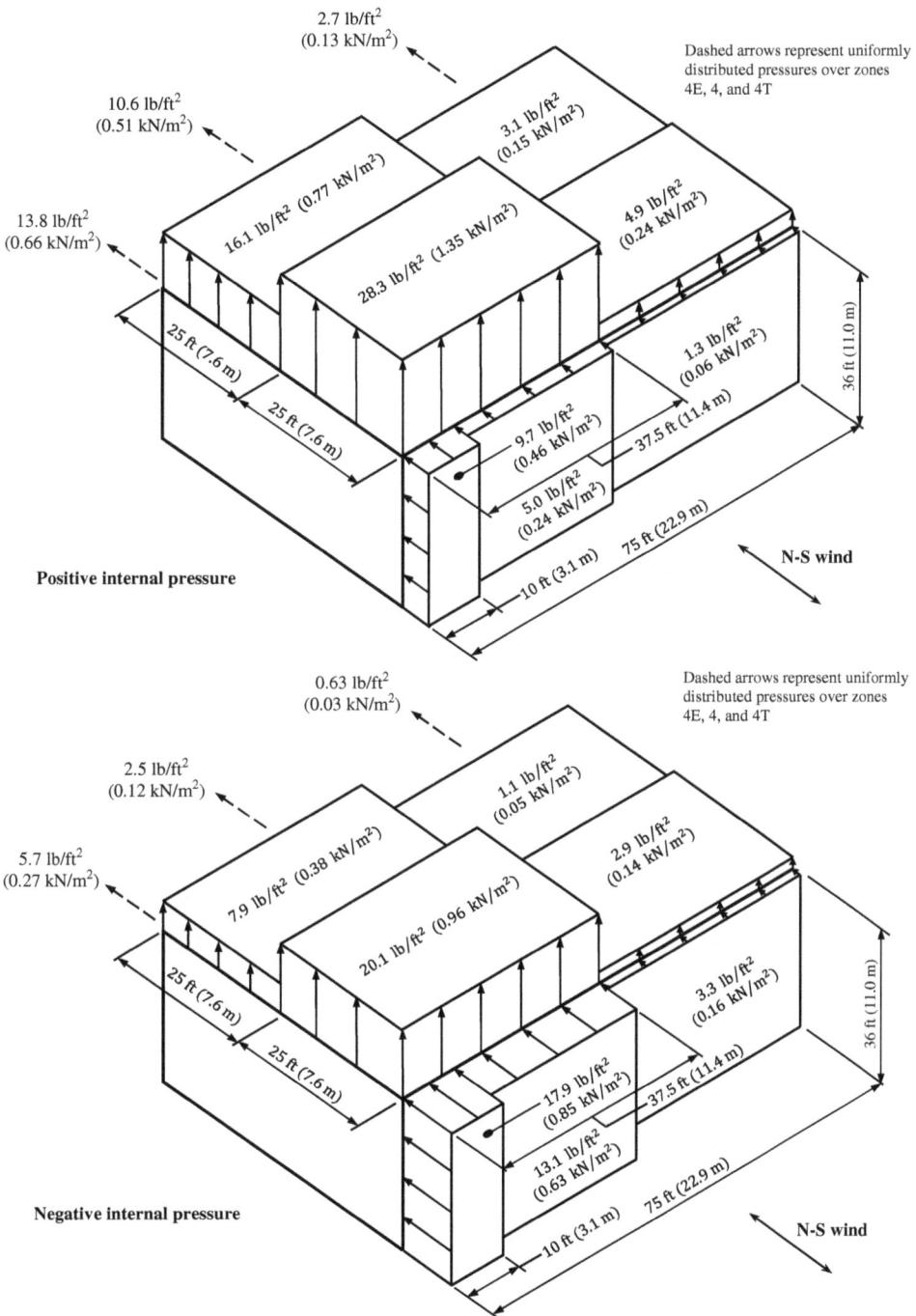

FIGURE 4.8 Design wind pressures including torsional loading in the N-S direction for the building in Example 4.1.

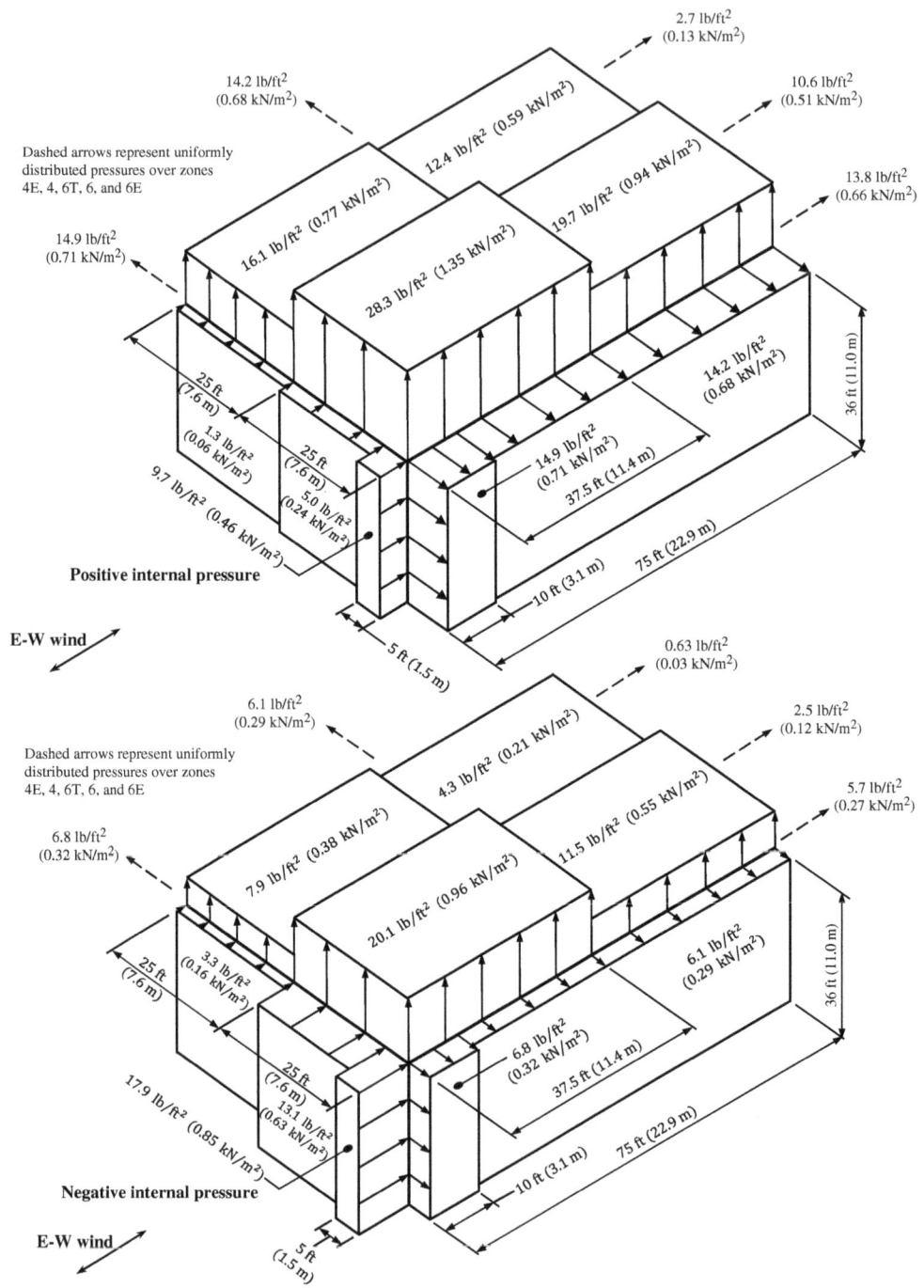

FIGURE 4.9 Design wind pressures including torsional loading in the E-W direction for the building in Example 4.1.

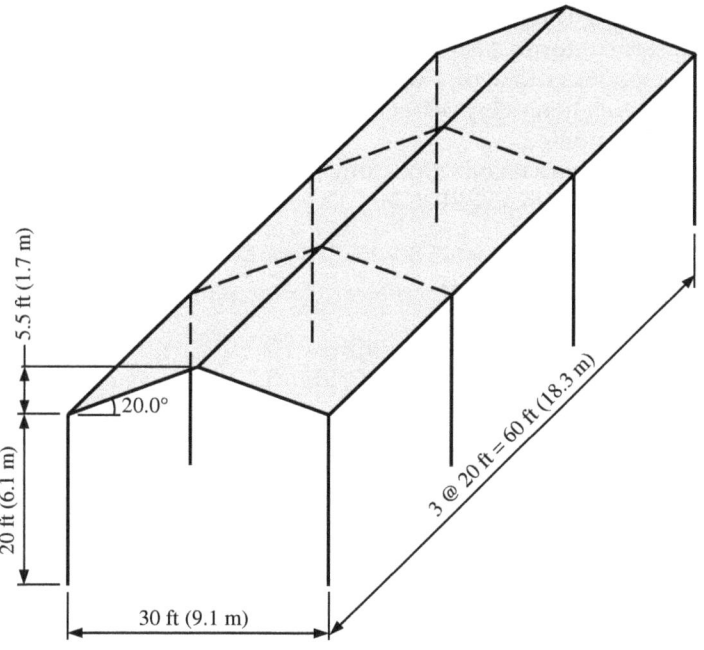

FIGURE 4.10 Open utility building in Example 4.2.

The minimum design wind loading prescribed in ASCE/SEI 28.3.4 must be considered as a load case in addition to the load cases above.

4.4.2 Example 4.2—Horizontal Wind Loads on an Open Utility Building with Transverse Frames and a Pitched Roof, MWFRS, Chapter 28, Part 1

Determine the horizontal wind pressure in the N-S direction (parallel to the ridge) for the open utility building in Fig. 4.10 using the requirements in Part 1 of Chapter 28 and the design data in Table 4.4.

Solution
Check if the building meets all the conditions in ASCE/SEI 28.1.2 so that Part 1 in Chapter 28 can be used to determine the design wind pressures on the MWFRS.

Location	Cedar Rapids, IA
Surface roughness	C
Topography	Not situated on a hill, ridge, or escarpment
Occupancy	Low risk to human life in the event of failure
Enclosure classification	Open (no walls)
Fundamental natural frequency of the MWFRS	>1 Hz

TABLE 4.4 Design Data for the Open Utility Building in Example 4.2

The building is regular-shaped and does not have any unusual geometric irregularities in spatial form. Also, the building does not have any response characteristics that make it subject to across-wind loading or similar effects, and it is not sited at a location where channeling effects or buffeting in the wake of upwind obstructions need to be considered.

The following two conditions must be checked to determine if the building is a low-rise building (ASCE/SEI 26.2):

- Mean roof height = $(25.5 + 20.0)/2 = 22.8$ ft (7.0 m) < 60 ft (18.3 m)
- Mean roof height = 22.8 ft (6.9 m) < least horizontal dimension = 30 ft (9.1 m)

Thus, the building is a low-rise building.

Therefore, the conditions in ASCE/SEI 28.1.2 are met and the requirements in Part 1 of Chapter 28 may be used.

The design procedure in Fig. 4.3 is used to determine the horizontal wind pressure in the N-S direction.

Step 1—Determine the velocity pressure, q_h, at the mean roof height   Fig. 2.3

The flowchart in Fig. 2.3 is used to determine q_h.

- Step 1a—Determine the surface roughness category   ASCE/SEI 26.7.2

  In Table 4.4, the surface roughness is given as C.

- Step 1b—Determine the exposure category   ASCE/SEI 26.7.3

  It is assumed that surface roughness C applies in all directions and that exposures B and D are not applicable. Therefore, the exposure category is C.

- Step 1c—Determine the terrain exposure constants   ASCE/SEI Table 26.11-1

  For Exposure C, $\alpha = 9.5$ and $z_g = 900$ ft (274.32 m).

- Step 1d—Determine the velocity pressure exposure coefficient, K_h
          ASCE/SEI Table 26.10-1

  At $z = 22.8$ ft (7.0 m):

$$K_h = 2.01(z/z_g)^{2/\alpha} = 2.01 \times (22.8/900)^{2/9.5} = 0.93$$

  In S.I.:

$$K_h = 2.01 \times (7.0/274.32)^{2/9.5} = 0.93$$

- Step 1e—Determine the topographic factor, K_{zt} ASCE/SEI 26.8

  Because the building is not located on a hill, ridge, or escarpment, $K_{zt} = 1.0$.

- Step 1f—Determine the wind directionality factor, K_d
          ASCE/SEI Table 26.6-1

  For the MWFRS of a building structure, $K_d = 0.85$.

- Step 1g—Determine the ground elevation factor, K_e ASCE/SEI 26.9

  Ground elevation factor can be taken as 1.0 for all elevations.

- Step 1h—Determine the risk category of the building
  ASCE/SEI Table 1.5-1

Due to the nature of its occupancy, this utility building falls under Risk Category I.

- Step 1i—Determine the basic wind speed, V Table 2.1

For Risk Category I, use IBC Figure 1609.3(4) or ASCE/SEI Figure 26.5-1A.

Equivalently, use Ref. 3 or Ref. 4 to obtain $V = 102$ mi/h (46 m/s) for Cedar Rapids, IA.

- Step 1j—Determine the wind velocity pressure, q_h ASCE/SEI 26.10.2

$$q_h = 0.00256 K_h K_{zt} K_d K_e V^2$$

$$= 0.00256 \times 0.93 \times 1.0 \times 0.85 \times 1.0 \times 102^2 = 21.1 \text{ lb/ft}^2$$

In S.I.:

$$q_h = 0.613 K_h K_{zt} K_d K_e V^2$$

$$= 0.613 \times 0.93 \times 1.0 \times 0.85 \times 1.0 \times 46^2 / 1,000 = 1.03 \text{ kN/m}^2$$

Step 2—Determine the external pressure coefficient, $(GC_{pf})_{windward}$
ASCE/SEI Figure 28.3-1

The external pressure coefficient for the windward side is determined from load case B in ASCE/SEI Figure 28.3-1. An average value of this coefficient is used on the windward face based on surfaces 5 and 5E:

$$(GC_{pf})_{windward} = (0.40 + 0.61)/2 = 0.51$$

Step 3—Determine the external pressure coefficient, $(GC_{pf})_{leeward}$
ASCE/SEI Figure 28.3-1

The external pressure coefficient for the leeward side is determined from load case B in ASCE/SEI Figure 28.3-1. An average value of this coefficient is used on the leeward face based on surfaces 6 and 6E:

$$(GC_{pf})_{windward} = [(-0.29) + (-0.43)]/2 = -0.36$$

Step 4—Determine the frame width factor, K_B ASCE/SEI 28.3.5

Because $B = 30$ ft (9.1 m) < 100 ft (30.5 m), K_B is equal to the following:

$$K_B = 1.8 - 0.01B = 1.8 - (0.01 \times 30.0) = 1.5$$

Step 5—Determine the solidity ratio, ϕ ASCE/SEI 28.3.5

Assume $\phi = A_S / A_E = 0.1$.

Step 6—Determine the shielding factor, K_S ASCE/SEI 28.3.5

$$K_S = 0.60 + 0.073(n - 3) + 1.25\phi^{1.8}$$

$$= 0.60 + [0.073 \times (4 - 3)] + (1.25 \times 0.1^{1.8}) = 0.69$$

Step 7—Determine the horizontal wind pressure, p   ASCE/SEI Equation (28.3-3)

$$p = q_h[(GC_{pf})_{windward} - (GC_{pf})_{leeward}]K_B K_S$$

$$= 21.1 \times [0.51 - (-0.36)] \times 1.5 \times 0.69 = 19.0 \text{ lb/ft}^2$$

In S.I.:

$$p = 1.01 \times [0.51 - (-0.36)] \times 1.5 \times 0.69 = 0.91 \text{ kN/m}^2$$

These wind pressures act in combination with the wind pressures on the roof determined by ASCE/SEI 27.3.2 (see Sec. 3.2.2 of this publication).

4.4.3   Example 4.3—Wind Pressures on a Retail Building, MWFRS, Chapter 28, Part 2

Determine the wind pressures in both directions on the MWFRS of the one-story retail building in Fig. 3.18 using the requirements in Part 2 of Chapter 28 and the design data in Table 3.7. Assume the angle of the roof is 5 degrees instead of 14 degrees and the floor and roof diaphragms are flexible, which transfer the lateral load effects to the concrete masonry walls.

Solution

Check if the building meets all the conditions in ASCE/SEI 28.1.2 and 28.5.2 so that Part 2 in Chapter 28 can be used to determine the design wind pressures on the MWFRS.

The building is regular-shaped and does not have any unusual geometric irregularities in spatial form. Also, the building does not have any response characteristics that make it subject to across-wind loading or similar effects, and it is not sited at a location where channeling effects or buffeting in the wake of upwind obstructions need to be considered.

The following conditions are also relevant (ASCE/SEI 28.5.2):

- The building is a simple diaphragm building as defined in ASCE/SEI 26.2 because windward and leeward wind loads are transmitted through the flexible diaphragms to the concrete masonry walls and there are no separations in the MWFRS.

- The building is a low-rise building as defined in ASCE/SEI 26.2 because the mean roof height = $(25.0 + 33.75)/2 = 29.4$ ft (9.0 m) < 60 ft (18.3 m) and the mean roof height = 29.4 ft (9.0 m) < least horizontal dimension = 125.0 ft (38.1 m).

- The building is enclosed.

- The building has been determined to be not flexible.

- The building has a symmetrical cross-section in each direction, and the slope of the roof is less than 45 degrees.

- The building is exempted from the torsional load cases indicated in note 5 of ASCE/SEI Figure 28.3-1 because the building is one-story with a mean roof height less than 30 ft (9.1 m).

Therefore, the conditions in ASCE/SEI 28.1.2 and 28.5.2 are met and the requirements in Part 2 of Chapter 28 may be used.

The design procedure in Fig. 4.5 is used to determine the design wind pressures in both directions.

Step 1—Determine the surface roughness category ASCE/SEI 26.7.2

The surface roughness category is given in the design data as B (see Table 3.7).

Step 2—Determine the exposure category ASCE/SEI 26.7.3

It is assumed that surface roughness B applies in all directions and that exposures C and D are not applicable. Therefore, the exposure category is B.

Step 3—Determine the topographic factor, K_{zt} ASCE/SEI 26.8

Because the building is not located on a hill, ridge, or escarpment, $K_{zt} = 1.0$.

Step 4—Determine the risk category of the building ASCE/SEI Table 1.5-1

Due to the nature of its occupancy, this retail building falls under Risk Category II.

Step 5—Determine the basic wind speed, V Table 2.1

For Risk Category II, use IBC Figure 1609.3(1) or ASCE/SEI Figure 26.5-1B. Equivalently, use Ref. 3 or Ref. 4 to obtain $V = 115$ mi/h (51 m/s) for Philadelphia, PA.

Step 6—Determine p_{s30} for zones A through H and the overhangs
ASCE/SEI Figure 28.5-1

Wind pressures, p_{s30}, for Exposure B and a mean roof height of 30 ft (9.1 m) are obtained directly from ASCE/SEI Figure 28.5-1 for $V = 115$ mi/h (51 m/s) and a roof slope of 5 degrees (see Table 4.5). Only load case 1 needs to be considered because of the roof slope.

Step 7—Determine the adjustment factor, λ, for height and exposure
ASCE/SEI Figure 28.5-1

For Exposure B and a mean roof height of 29.4 ft (9.0 m), $\lambda = 1.0$.

Step 8—Determine the net design wind pressures, p_s ASCE/SEI Equation (28.5-1)

$$p_s = \lambda K_{zt} p_{s30} = 1.0 \times 1.0 \times p_{s30} = p_{s30}$$

Therefore, the net design wind pressures in Table 4.5 are applied to the surfaces of the building in accordance with cases A and B in ASCE/SEI Figure 28.5-1.

According to note 7 in ASCE/SEI Figure 28.5-1, the total horizontal load must not be less than that determined by assuming $p_s = 0$ in zones B and D. Because the net pressures in zones B and D act in the direction opposite to

Horizontal Pressures, lb/ft² (kN/m²)				Vertical Pressures, lb/ft² (kN/m²)				Overhangs, lb/ft² (kN/m²)	
A	B	C	D	E	F	G	H	E_{OH}	G_{OH}
21.0	−10.9	13.9	−6.5	−25.2	−14.3	−17.5	−11.1	−35.3	−27.6
(1.01)	(−0.52)	(0.67)	(−0.31)	(−1.21)	(−0.69)	(−0.84)	(−0.53)	(−1.69)	(−1.32)

TABLE 4.5 Wind Pressures, p_{s30}, on the MWFRS of the Building in Example 4.3

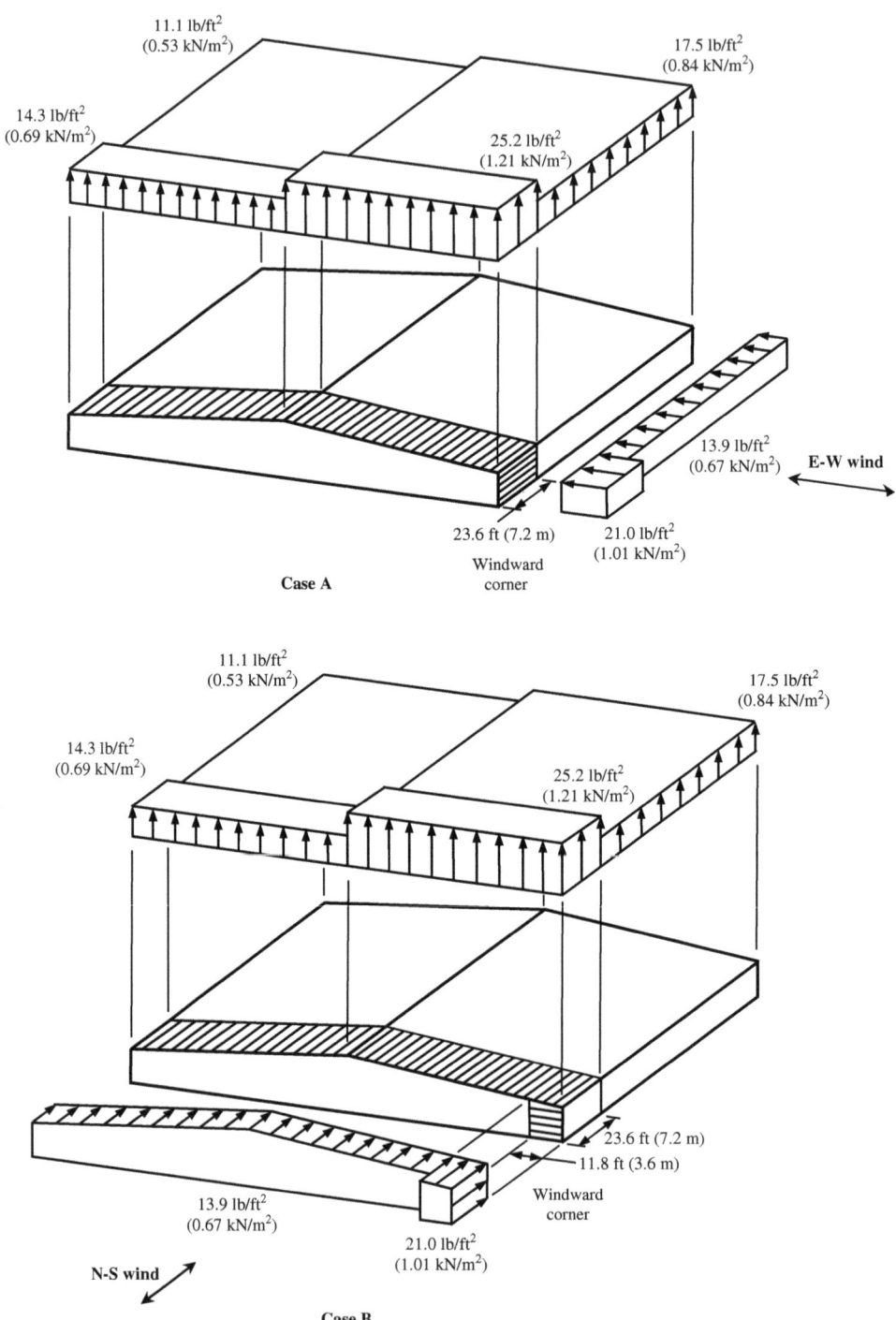

FIGURE 4.11 Design wind pressures on the MWFRS of the building in Example 4.3.

those in zones A and C, these pressures decrease the horizontal load. Therefore, the pressures in zones B and D are set equal to zero when analyzing the building for wind in the E-W direction (case A).

The load patterns in cases A and B are to be applied to each corner of the building, that is, each corner must be considered a windward corner (see note 2 in ASCE/SEI Figure 28.5-1). Eight different load cases must be examined (four in case A and four in case B). One load pattern for wind in the E-W direction (case A) and one for wind in the N-S direction (case B) are shown in Fig. 4.11.

The distance a is equal to the following:

$a = $ smaller of

$$\begin{cases} 0.1 \times \text{least horizontal dimension} = 0.1 \times 125 = 12.5 \text{ ft (3.8 m)} \\ 0.4h = 0.4 \times 29.4 = 11.8 \text{ ft (3.6 m)} \end{cases}$$

Minimum $a = $ greater of

$$\begin{cases} 0.04 \times \text{least horizontal dimension} = 0.04 \times 125 = 5 \text{ ft (1.5 m)} \\ 3 \text{ ft (0.91 m)} \end{cases}$$

Therefore, $a = 11.8$ ft (3.6 m).

The minimum design wind load case of ASCE/SEI 28.5.4 must also be considered. The load effects from the design wind pressures in Table 4.5 must not be less than the load effects assuming $p_s = +16$ lb/ft² (0.77 kN/m²) in zones A and C, $p_s = +8$ lb/ft² (0.38 kN/m²) in zones B and D, and $p_s = 0$ in zones E through H.

CHAPTER 5

Wind Loads on Building Appurtenances and Other Structures: MWFRS (Directional Procedure)

5.1 Overview

This chapter contains the requirements in ASCE/SEI Chapter 29 for determining wind pressures and loads on the MWFRSs of the following:

- Building appurtenances, including rooftop structures and rooftop equipment.
- Other structures of all heights, including solid freestanding walls, solid freestanding signs, chimneys, tanks, open signs, single-plane open frames, and trussed towers.

The provisions in Chapter 29 are applicable to appurtenances or structures that comply with the following (ASCE/SEI 29.1.2):

- The structure is regular-shaped, that is, the structure has no unusual geometrical irregularities in spatial form.
- The structure does not have response characteristics that make it subject to across-wind loading, vortex shedding, or instability caused by galloping or flutter. Additionally, the structure is not located at a site where channeling effects or buffeting in the wake of upwind obstructions warrant special consideration.

Structures not meeting these conditions must be designed by either recognized literature that documents such wind load effects or by the wind tunnel procedure in ASCE/SEI Chapter 31 (ASCE/SEI 29.1.3).

Reduction in wind pressure due to apparent shielding by surrounding buildings, other structures, or terrain features is not permitted (ASCE/SEI 29.1.4). Such shielding may be modified or completely removed during the lifespan of the structure, which could result in significantly higher wind loads.

5.2 Solid Freestanding Walls and Solid Signs

5.2.1 Design Wind Forces for Solid Freestanding Walls and Solid Freestanding Signs

The design wind force, F, for the MWFRS of solid freestanding walls and solid freestanding signs is determined by ASCE/SEI Equation (29.3-1) (walls and signs with openings comprising less than 30 percent of the gross area are classified as solid; for walls or signs with openings greater than or equal to 30 percent, the wind force must be determined in accordance with ASCE/SEI 29.4, which is covered in Sec. 5.3 of this publication):

$$F = q_h GC_f A_s \tag{5.1}$$

The velocity pressure at the mean roof height of the wall or sign, q_h, is determined in accordance with ASCE/SEI 26.10 (see Sec. 2.7 of this publication) and the gust-effect factor, G, is determined by ASCE/SEI 26.11 (see Sec. 2.8 of this publication). Net force coefficients, C_f, are given in ASCE/SEI Figure 29.3-1 as a function of the geometrical properties of the wall or sign. The term A_s is the gross area of the solid freestanding wall or sign.

In general, three cases (A, B, and C) must be investigated where F is applied at different locations on the sign or wall (see ASCE/SEI Figure 29.3-1 and Table 5.1). Cases A and B must always be considered and case C is applicable where the width of the sign or wall, B, is greater than or equal to 2 times the vertical dimension of the sign or wall, s.

A flowchart to determine the design wind force for solid freestanding walls and solid freestanding signs is given in Fig. 5.1 (see also Fig. 5.2).

5.2.2   Solid Attached Signs

The components and cladding (C&C) requirements for walls in ASCE/SEI Chapter 30 must be used to determine wind pressures on solid signs attached to the wall of a building provided the plane of the sign is parallel to and in contact with the plane of the wall and the sign does not extend beyond the side or top edges of the wall (see Chap. 6 of this publication). In such cases, the internal pressure coefficient, (GC_{pi}), is set equal to zero.

Load Case	Location of Resultant Force, *F*	
	s/h < 1	*s/h* = 1
A	*F* acts normal to the face of the wall or sign through the geometric center	*F* acts normal to the face of the wall or sign at a distance above the geometric center equal to 0.05 times the average height of the wall or sign
B	*F* acts normal to the face of the wall or sign at a distance from the geometric center toward the windward edge equal to 0.2 times the average width of the wall or sign	*F* acts normal to the face of the wall or sign at a distance from the geometric center toward the windward edge equal to 0.2 times the average width of the wall or sign and a distance above the geometric center equal to 0.05 times the average height of the wall or sign
C	*F* acts normal to the face of the wall or sign through the geometric centers of each region identified in ASCE/SEI Figure 29.3-1	*F* acts normal to the face of the wall or sign through the geometric centers of each region identified in ASCE/SEI Figure 29.3-1 and at a distance above the geometric center equal to 0.05 times the average height of the wall or sign

TABLE 5.1 Load Cases for Solid Freestanding Walls and Solid Signs

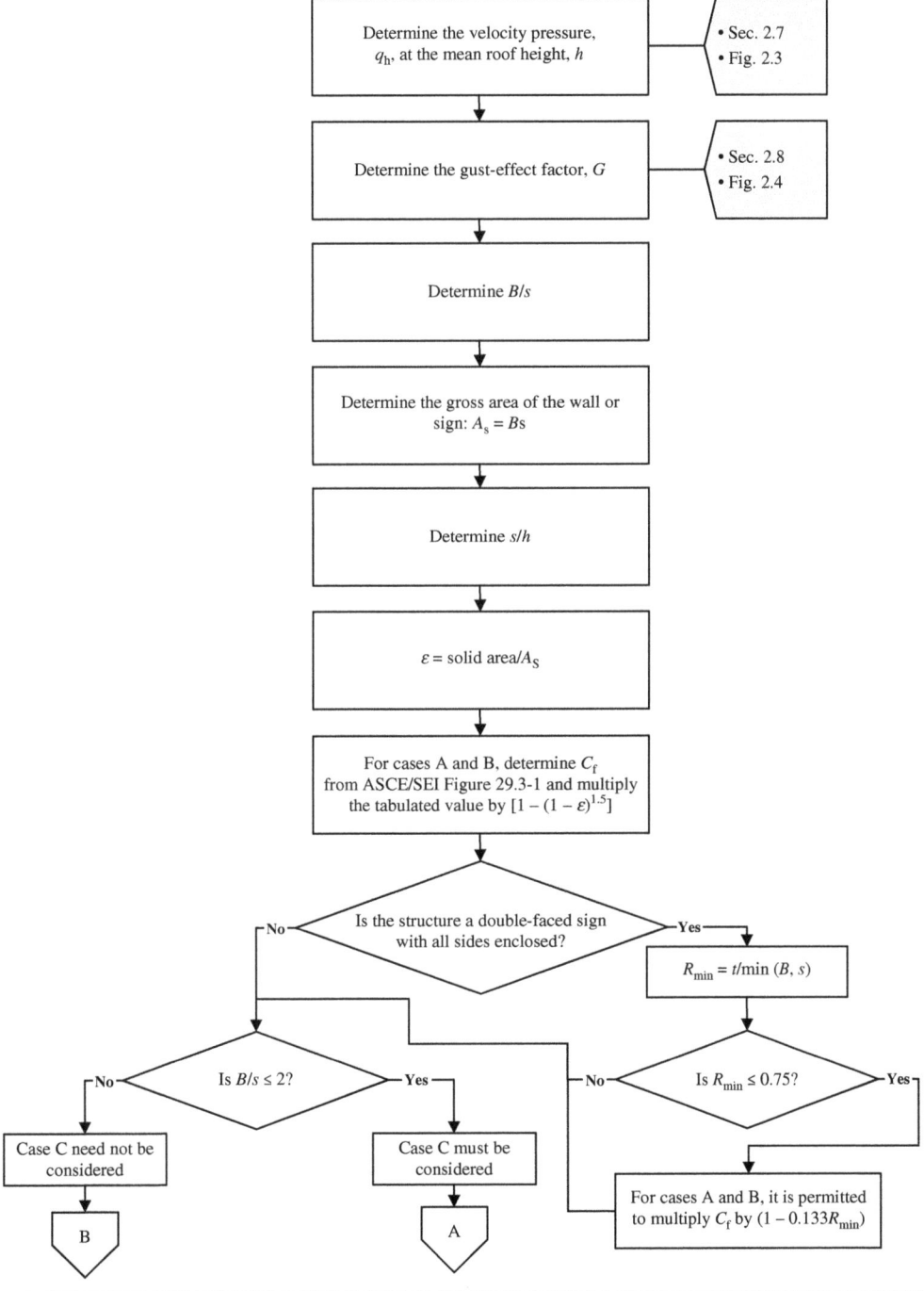

FIGURE 5.1 Design wind forces for solid freestanding walls and solid freestanding signs.

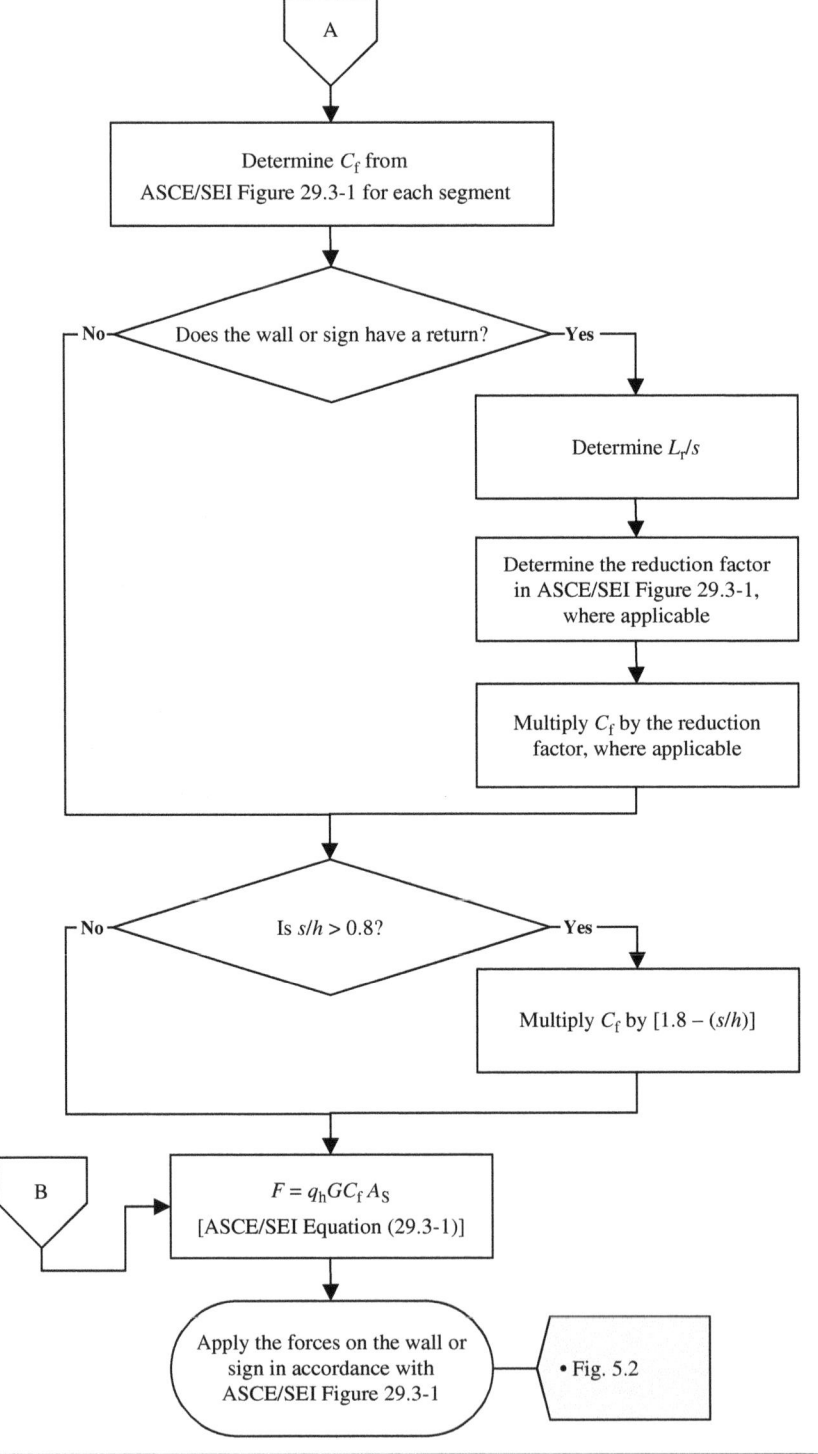

Figure 5.1 (Continued)

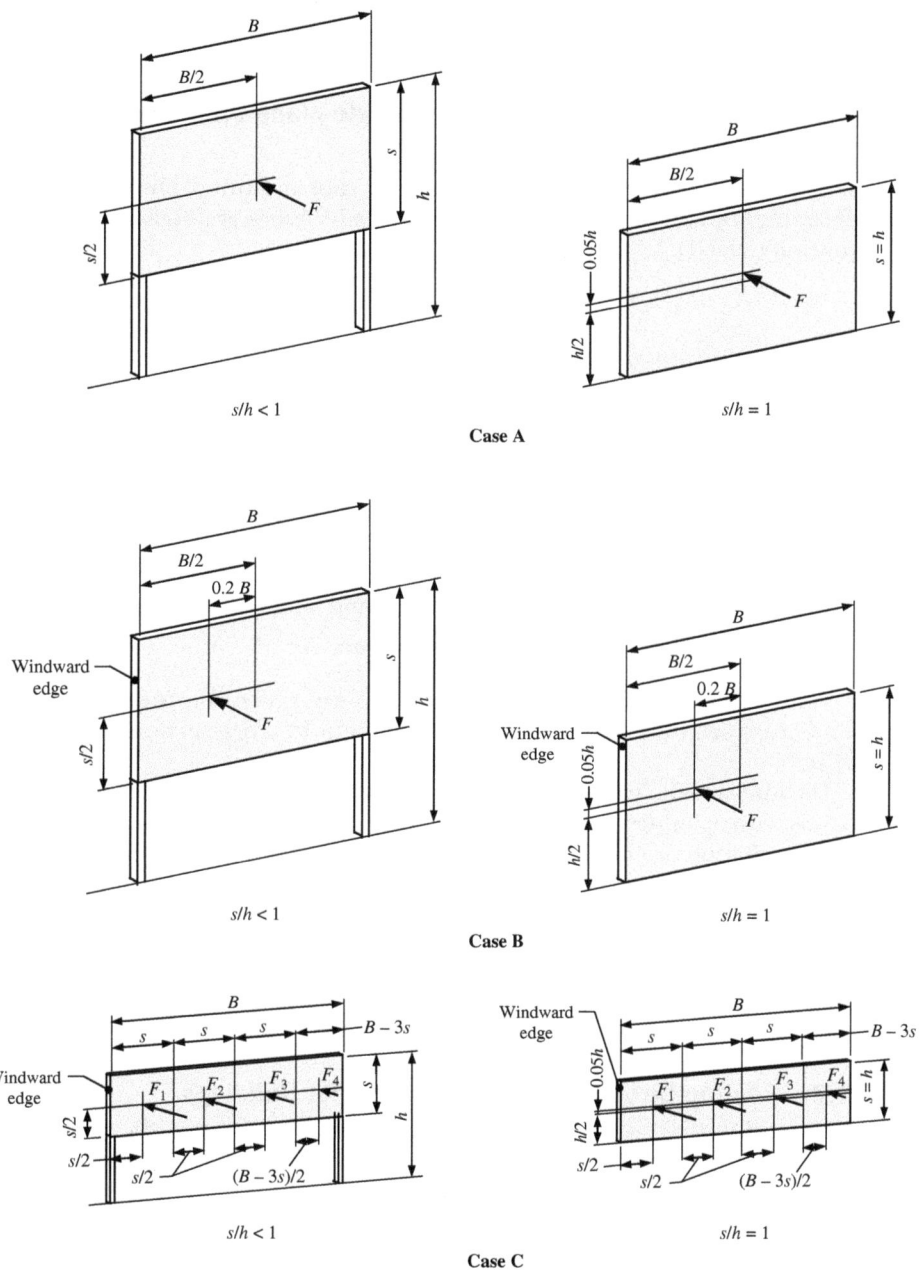

FIGURE 5.2   Cases A, B, and C for solid freestanding walls and solid freestanding signs.

This procedure is also applicable to signs attached to but not in direct contact with the wall provided the gap between the sign and wall is no more than 3 ft (0.91 m) and the edge of the sign is at least 3 ft (0.91 m) from the free edges of the wall (that is, side and top edges and bottom edges of elevated walls).

5.3   Other Structures

5.3.1   Chimneys, Tanks, Open Signs, Single-plane Open Frames, and Trussed Towers

The design wind force, F, for ground or roof-mounted chimneys, tanks, open signs, single-plane open frames, and trussed towers is determined by ASCE/SEI Equation (29.4-1):

$$F = q_z G C_f A_f \qquad (5.2)$$

The velocity pressure, q_z, at height z of the centroid of the area, A_f, is determined in accordance with ASCE/SEI 26.10 (see Sec. 2.7 of this publication) and the gust-effect factor, G, is determined by ASCE/SEI 26.11 (see Sec. 2.8 of this publication). The area A_f is the projected area of the structure normal to the wind except where the force coefficient, C_f, is specified for the actual surface area.

Force coefficients, C_f, are given in the following figures:

- ASCE/SEI Figure 29.4-1: Chimneys, tanks, and similar structures
- ASCE/SEI Figure 29.4-2: Open signs and single-plane open frames
- ASCE/SEI Figure 29.4-3: Trussed towers

Values of C_f in ASCE/SEI Figure 29.4-1 are given for square, hexagonal, and round cross-sections as a function of the height to cross-sectional dimension of the section.

The force coefficients in ASCE/SEI Figure 29.4-2 are applicable to open signs, that is, signs with openings comprising 30 percent or more of the gross area. Signs not meeting this criterion are classified as solid signs and the force coefficients in ASCE/SEI Figure 29.3-1 must be used (see Sec. 5.2.1 of this publication).

The force coefficients in ASCE/SEI Figure 29.4-3 are for trussed towers with square and triangular cross-sections.

A flowchart to determine the design wind force for other structures is given in Fig. 5.3.

5.3.2   Rooftop Structures and Equipment for Buildings

Lateral and vertical wind forces must be determined in accordance with ASCE/SEI 29.4.1 for structures and equipment on the rooftop of buildings (excluding roof-mounted solar panels, which are covered in ASCE/SEI 29.4.3 and 29.4.4; see Secs. 5.3.4 and 5.3.5 of this publication).

Lateral force, F_h, and uplift force, F_v, are determined by ASCE/SEI Equations (29.4-2) and (29.4-3), respectively (see Table 5.2 and Fig. 5.4). Forces F_h and F_v are applied at the centroids of the areas A_f and A_r, respectively.

The basic wind speed, V, used in determining q_h at the mean roof height of the building must be based on the greater of the following risk categories (ASCE/SEI 26.10.2): (1) the risk category of the building on which the structure or equipment is located or (2) the risk category for any facility to which the structure or equipment provides a necessary service.

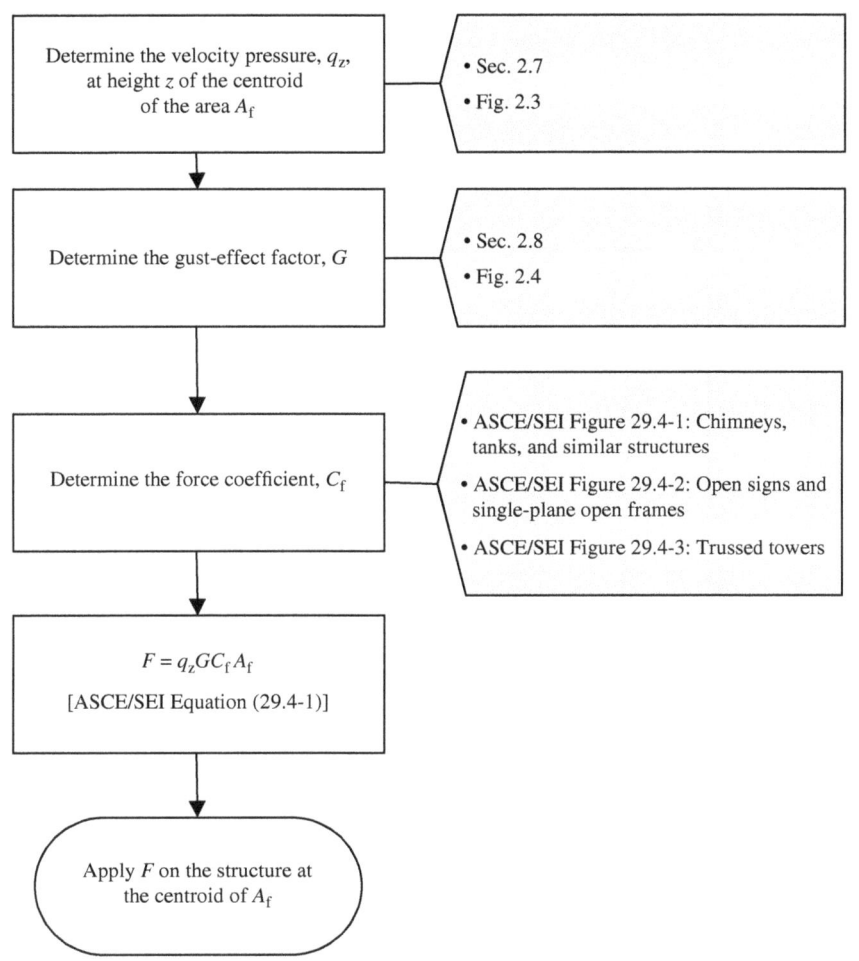

FIGURE 5.3 Flowchart to determine the design wind forces for other structures.

Lateral Force, F_h	Uplift Force, F_v
$F_h = q_h(GC_r)A_f$	$F_v = q_h(GC_r)A_r$
A_f = vertical projected area of the rooftop structure or equipment on a plane normal to the direction of wind	A_r = horizontal projected area of the rooftop structure or equipment
$(GC_r) = \begin{cases} 1.9 \text{ for } A_f < 0.1\ Bh \\ 2.0 - (A_f/Bh) \text{ for } 0.1\ Bh \le A_f \le Bh \end{cases}$	$(GC_r) = \begin{cases} 1.5 \text{ for } A_r < 0.1\ BL \\ 1.0 + 0.56[1 - (A_r/BL)] \text{ for } 0.1BL \le A_r \le BL \end{cases}$

q_h = Velocity pressure evaluated at the mean roof height of the building

TABLE 5.2 Lateral and Uplift Forces on Rooftop Structures and Equipment for Buildings

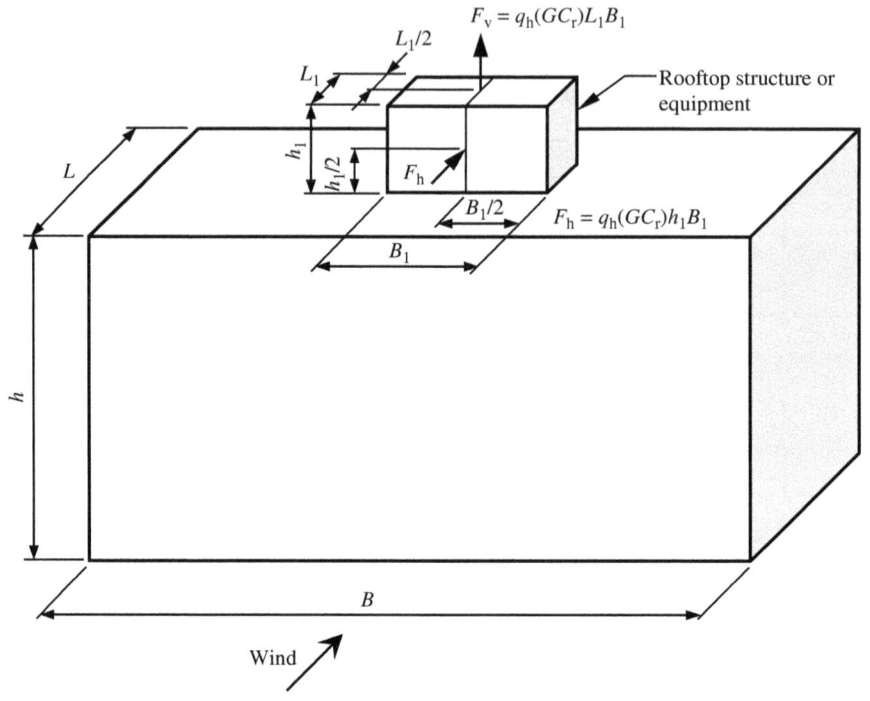

Figure 5.4 Lateral and uplift wind forces on rooftop structures and equipment.

5.3.3 Circular Bins, Silos, and Tanks with $h \leq 120$ ft (36.6 m), $D \leq 120$ ft (36.6 m) and $0.25 \leq H/D \leq 4$

Design wind loads for circular bins, silos, and tanks with a mean roof height, h, less than or equal to 120 ft (36.6 m), a diameter, D, less than or equal to 120 ft (36.6 m), and a ratio of solid cylinder height, H, to diameter, D, in the range of 0.25 and 4 are determined in accordance with ASCE/SEI 29.4.2.

A summary of the wind loads and pressures for the external walls, roof, and, where applicable, the undersides of isolated circular bins, silos, and tanks is given in Table 5.3 (see Fig. 5.5). Isolated structures are defined as structures of similar size with a center-to-center spacing greater than $2D$.

Where the center-to-center spacing of three or more structures is less than $1.25D$, the structures are assumed to be grouped and the wind pressures are determined in accordance with ASCE/SEI 29.4.2.4.

A summary of the wind loads and pressures for grouped circular bins, silos, and tanks is given in Table 5.4.

For similar structures with a center-to-center spacing less than $2D$ and greater than $1.25D$, it is permitted to determine wind loads and pressures using linear interpolation of the force and pressure coefficients presented above (ASCE/SEI 29.4.2).

The flowchart in Fig. 5.6 can be used to determine the wind forces and pressures on isolated or grouped circular bins, silos, and tanks.

Element	Design Wind Load (F) or Pressure (p)
External walls	$$F = 0.63q_zGDH$$
Roofs	$$p = q_h[GC_p - (GC_{pi})]$$ where G = gust-effect factor based on the natural frequency of the entire structure (that is, the supported structure and the support structure) [ASCE/SEI 26.11] C_p = external pressure coefficient for zones 1 and 2 in ASCE/SEI Figure 29.4-5 for conical, flat, or dome roofs with an angle less than 10 degrees and for conical roofs with an angle between 10 and 30 degrees, inclusive* (GC_{pi}) = internal pressure coefficient for roofed structures (ASCE/SEI 26.11)
Undersides	$$p = q_h[GC_p - (GC_{pi})]$$ where $C_p = \begin{cases} 0.8, -0.6 \text{ for } H/3 < C \leq H \\ 0.8(C/h), -0.6(C/h) \text{ for } C \leq H/3 \end{cases}$

*For domed roofs with a roof angle greater than 10 degrees, the external pressures must be determined from ASCE/SEI Figure 27.3-2.

TABLE 5.3 Design Wind Loads and Pressures for Isolated Circular Bins, Silos, and Tanks

5.3.4 Rooftop Solar Panels for Buildings of All Heights with Flat Roofs or Gable or Hip Roofs with Slopes Less than 7 Degrees

Provisions to determine design wind pressures on rooftop solar panels on enclosed or partially enclosed buildings of all heights with flat roofs or with gable or hip roofs with a roof slope less than or equal to 7 degrees are given in ASCE/SEI 29.4.3. These provisions are applicable where the following criteria are satisfied (see ASCE/SEI Figure 29.4-7):

- Panel chord length, $L_p \leq 6.7$ ft (2.0 m)
- Angle the panel makes with the roof surface, $\omega \leq 35$ degrees
- Height of the panel above the roof at the lower edge of the panel, $h_1 \leq 2$ ft (0.61 m)
- Height of the panel above the roof at the upper edge of the panel, $h_2 \leq 4$ ft (1.2 m)
- Minimum gap between all panels = 0.25 in. (6.4 mm)
- Spacing of gaps between panels ≤ 6.7 ft (2.0 m)
- Minimum horizontal clear distance between the panels and the edge of the roof

$$\geq \text{larger of} \begin{cases} 2(h_2 - h_{pt}) \\ 4 \text{ ft (1.2 m)} \end{cases}$$

The term h_{pt} is the mean parapet height above the adjacent roof surface.

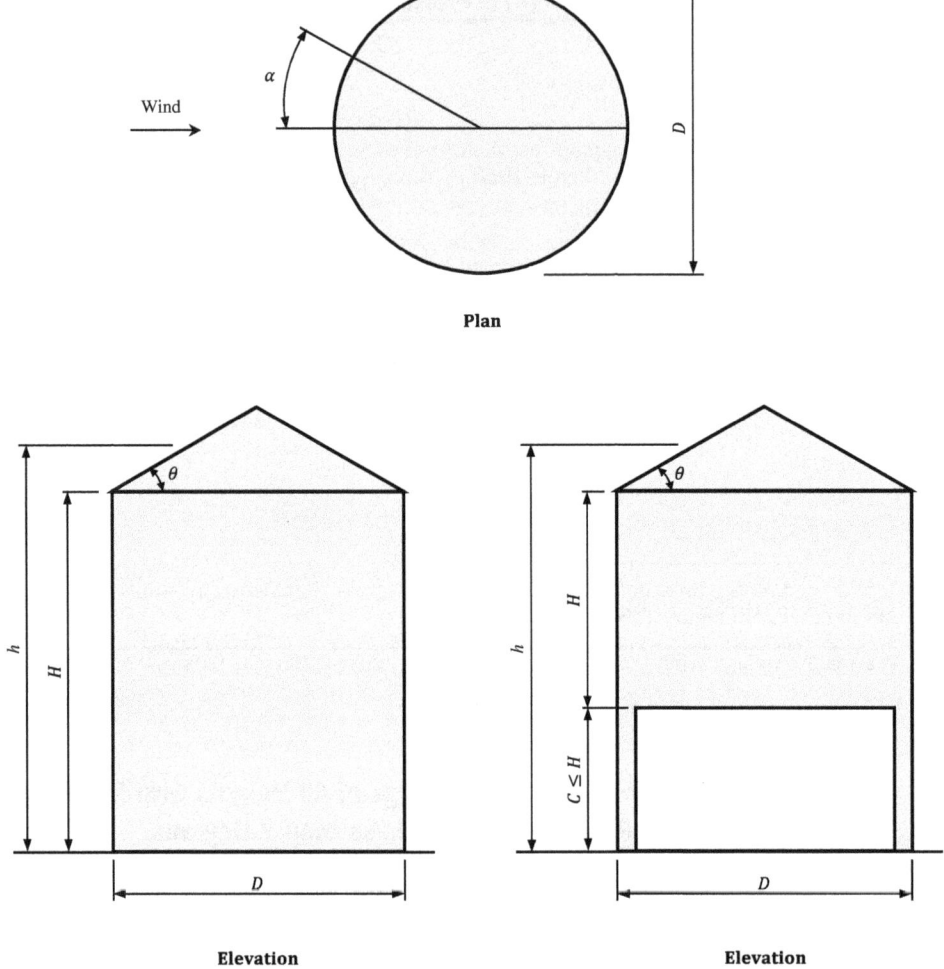

Plan

Elevation   **Elevation**

FIGURE 5.5   Plan and elevations of circular bins, silos, and tanks on the ground or supported by columns.

The design wind pressure, p, for rooftop solar panels is determined by ASCE/SEI Equation (29.4-5):

$$p = q_h(GC_m) \qquad (5.3)$$

In this equation, q_h is the velocity pressure determined by ASCE/SEI Equation (26.10-1) at the mean roof height of the building and (GC_m) is the net pressure coefficient for rooftop solar panels determined by ASCE/SEI Equation (29.4-6):

$$(GC_m) = \gamma_p \gamma_c \gamma_E (GC_m)_{nom} \qquad (5.4)$$

The terms in Eq. (5.4) are defined in Table 5.5.

Element	Design Wind Load (*F*) or Pressure (*p*)
External walls	$$F = q_z GC_f DH$$ where C_f is determined by ASCE/SEI Figure 29.4-6
Roofs	$$p = q_h[GC_p - (GC_{pi})]$$ where G = gust-effect factor based on the natural frequency of the entire structure (that is, the supported structure and the support structure) (ASCE/SEI 26.11) C_p = external pressure coefficient for zones 1 and 2 in ASCE/SEI Figure 29.4-6 for conical, flat, or dome roofs with an angle less than 10 degrees and for conical roofs with an angle between 10 and 30 degrees, inclusive* (GC_{pi}) = internal pressure coefficient for roofed structures (ASCE/SEI 26.11)

---

*For domed roofs with a roof angle greater than 10 degrees, the external pressures must be determined from ASCE/SEI Figure 27.3-2.

TABLE 5.4 Design Wind Loads and Pressures for Grouped Circular Bins, Silos, and Tanks

Equations to determine $(GC_{rn})_{nom}$ are given in Table 5.6 (see Ref. 5). Linear interpolation is permitted where ω is between 5 and 15 degrees.

The flowchart in Fig. 5.7 can be used to determine the design wind pressure on rooftop solar panels for enclosed and partially enclosed buildings with a roof slope less than or equal to 7 degrees.

In lieu of the procedure outlined above, the provisions in ASCE/SEI 29.4.4 (see Sec. 5.3.5 of this publication) are permitted to be used to determine wind pressures on rooftop solar panels provided all the following conditions are met (ASCE/SEI 29.4.3):

- $\omega \le 2$ degrees
- $h_2 \le 0.83$ ft (0.25 m)
- Minimum gap between all panels ≥ 0.25 in. (6.4 mm)
- Spacing of gaps between panels ≤ 6.7 ft (2.0 m)

Roof structures supporting solar panels must be designed for the maximum effects due to the two load cases in Table 5.7.

5.3.5 Rooftop Solar Panels Parallel to the Roof Surface on Buildings of All Heights and Roof Slopes

Provisions to determine design wind pressures for rooftop solar panels parallel to the roof surface on enclosed and partially enclosed buildings of all heights and roof slopes are given in ASCE/SEI 29.4.4. These provisions are applicable where the following criteria are satisfied:

- Panels are parallel to the roof surface within a tolerance of 2 degrees
- Maximum height of the panel above the roof surface $h_2 \le 0.83$ ft (0.25 m)
- Minimum gap between all panels ≥ 0.25 in. (6.4 mm)

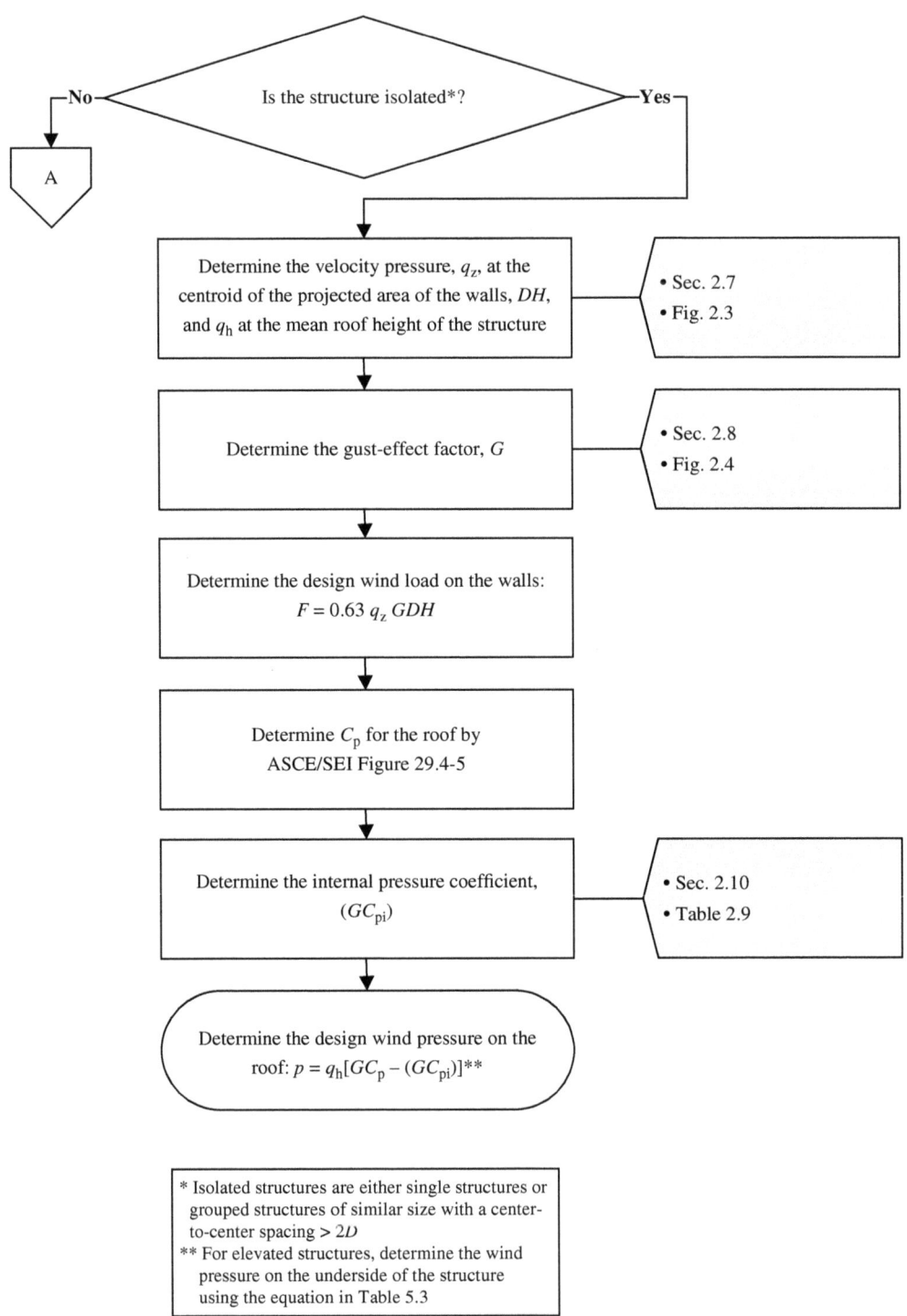

FIGURE 5.6   Flowchart to determine the wind forces and pressures on isolated or grouped circular bins, silos, and tanks.

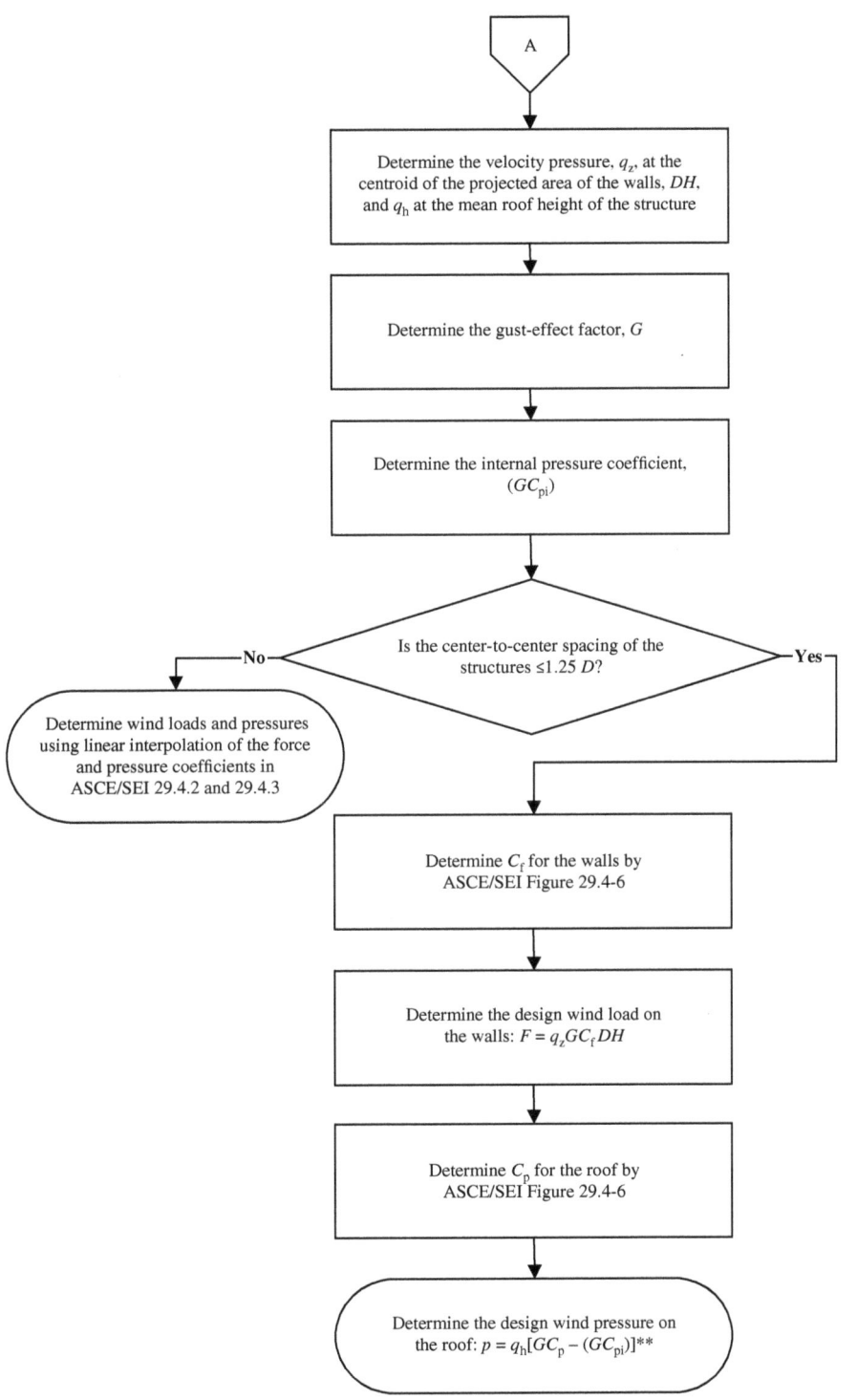

Figure 5.6 *(Continued)*

Parapet height factor, γ_p	$\gamma_p = \text{minimum of} \begin{cases} 1.2 \\ 0.9 + (h_{pt}/h) \end{cases}$
Panel chord factor, γ_c	$\gamma_c = \text{maximum of} \begin{cases} 0.6 + 0.06\, L_p \\ 0.8 \end{cases}$
Array edge factor, γ_E	$\gamma_E = \begin{cases} \text{1.5 for uplift loads on exposed panels and within a} \\ \text{distance of 1.5 } L_p \text{ from the end of a row at an} \\ \text{exposed edge of the array*} \\ \\ \text{1.0 elsewhere for uplift loads and for all downward} \\ \text{loads} \end{cases}$
Net nominal pressure coefficient, $(GC_{rn})_{nom}$	$(GC_{rn})_{nom}$ is determined from ASCE/SEI Figure 29.4-7 based on ω and the normalized wind area, A_n, corresponding to interior, edge, and corner conditions (identified as zones 1, 2, and 3 in ASCE/SEI Figure 29.4-7, respectively)**

*A panel is defined as exposed if the distance to the roof edge $d_1 > 0.5h$ and one of the following applies: (1) distance to the adjacent array $d_1 > \max\,[4h_2, 4\text{ ft } (1.2\text{ m})]$ or (2) distance to the next adjacent panel $d_2 > \max\,[4h_2, 4\text{ ft } (1.2\text{ m})]$.

**$A_n = 1{,}000A/\max\,[L_b, 15\text{ ft } (4.6\text{ m})]^2$ where A is the effective wind area of the structural element of the solar panel being considered and L_b is the normalized building length, which is equal to the minimum of the following: (1) $0.4(hW_L)^{0.5}$, (2) h, or (3) W_S where W_L and W_S are the widths of the long and short sides of the building, respectively (see ASCE/Figure 29.4-7).

TABLE 5.5   Net Pressure Coefficient for Rooftop Solar Panels, (GC_{rn})

ω (deg)	A_n	Zone 1	Zone 2	Zone 3
0 to 5	1 to 500	$-0.4261\log_{10}(A_n) + 1.500$	$-0.5743\log_{10}(A_n) + 2.000$	$-0.6669\log_{10}(A_n) + 2.300$
	500 to 5,000	$-0.2500\log_{10}(A_n) + 1.025$	$-0.3000\log_{10}(A_n) + 1.260$	$-0.3500\log_{10}(A_n) + 1.445$
15 to 35	1 to 500	$-0.5372\log_{10}(A_n) + 2.000$	$-0.8337\log_{10}(A_n) + 2.900$	$-1.0004\log_{10}(A_n) + 3.500$
	500 to 5,000	$-0.2500\log_{10}(A_n) + 1.225$	$-0.2500\log_{10}(A_n) + 1.325$	$-0.3000\log_{10}(A_n) + 1.610$

TABLE 5.6   Equations to Determine $(GC_{rn})_{nom}$

- Spacing of gaps between panels ≤ 6.7 ft (2.0 m)
- Distance from the edge of the panel array to the roof edge, gable ridge, or hip ridge $\geq 2h_2$

The design wind pressure, p, is determined by ASCE/SEI Equation (29.4-7):

$$p = q_h(GC_p)\gamma_E\gamma_a \tag{5.5}$$

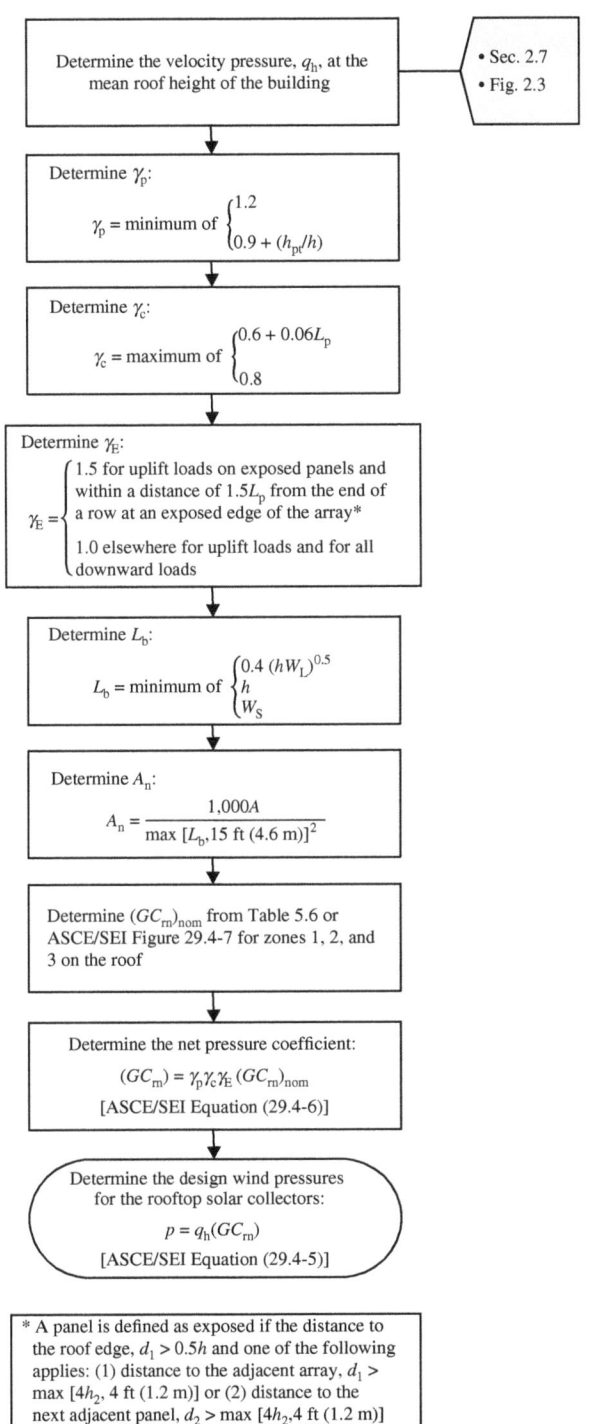

Determine the velocity pressure, q_h, at the mean roof height of the building

- Sec. 2.7
- Fig. 2.3

Determine γ_p:

$$\gamma_p = \text{minimum of} \begin{cases} 1.2 \\ 0.9 + (h_{pt}/h) \end{cases}$$

Determine γ_c:

$$\gamma_c = \text{maximum of} \begin{cases} 0.6 + 0.06L_p \\ 0.8 \end{cases}$$

Determine γ_E:

$$\gamma_E = \begin{cases} 1.5 \text{ for uplift loads on exposed panels and within a distance of } 1.5L_p \text{ from the end of a row at an exposed edge of the array*} \\ 1.0 \text{ elsewhere for uplift loads and for all downward loads} \end{cases}$$

Determine L_b:

$$L_b = \text{minimum of} \begin{cases} 0.4\,(hW_L)^{0.5} \\ h \\ W_S \end{cases}$$

Determine A_n:

$$A_n = \frac{1,000A}{\max\,[L_b, 15\text{ ft }(4.6\text{ m})]^2}$$

Determine $(GC_{rn})_{nom}$ from Table 5.6 or ASCE/SEI Figure 29.4-7 for zones 1, 2, and 3 on the roof

Determine the net pressure coefficient:

$$(GC_{rn}) = \gamma_p \gamma_c \gamma_E\,(GC_{rn})_{nom}$$

[ASCE/SEI Equation (29.4-6)]

Determine the design wind pressures for the rooftop solar collectors:

$$p = q_h(GC_{rn})$$

[ASCE/SEI Equation (29.4-5)]

* A panel is defined as exposed if the distance to the roof edge, $d_1 > 0.5h$ and one of the following applies: (1) distance to the adjacent array, $d_1 > \max\,[4h_2, 4\text{ ft }(1.2\text{ m})]$ or (2) distance to the next adjacent panel, $d_2 > \max\,[4h_2, 4\text{ ft }(1.2\text{ m})]$

FIGURE 5.7 Flowchart to determine the design wind pressures on rooftop solar panels for enclosed and partially enclosed buildings with a roof slope less than or equal to 7 degrees.

Solar panels are present	• Wind pressures on solar panels need not be applied simultaneously to the roof C&C wind pressures determined in accordance with ASCE/SEI Chapter 30 for the roof area covered by the solar panels. • Roof structural members beneath the solar panels must be designed for the reactions from wind pressure on the solar panels only. • Roof structural members partially covered by solar panels must be designed for both the solar panel and the roof C&C wind pressures applied over the applicable lengths of the member.
Solar panels are not present	Roof structural members must be designed for the roof C&C wind pressures assuming the solar panels may be removed in the future.

TABLE 5.7 Load Cases for Roof Structures Supporting Solar Panels

In this equation, q_h is the velocity pressure determined by ASCE/SEI Equation (26.10-1) at the mean roof height of the building and (GC_p) are the net pressure coefficients for C&C of roofs determined from ASCE/SEI Figures 30.3-2A-I through 30.3-7 or ASCE/SEI Figure 30.5-1.

Values of the array edge factor, γ_E, are defined in ASCE/SEI 29.4.3 (see Table 5.5). Panels are defined as being exposed where the distance d_1 from the panel edge to the roof edge is greater than $0.5h$ and one of the following applies:

- d_1 = distance to the adjacent solar array > 4 ft (1.2 m)
- d_2 = distance to the next adjacent panel > 4 ft (1.2 m)

The solar panel pressure equalization factor, γ_a, is defined in ASCE/SEI Figure 29.4-8 based on the effective wind area, A.

The two load cases in Table 5.7 must be considered for roof structures supporting solar panels governed by ASCE/SEI 29.4.4.

The flowchart in Fig. 5.8 can be used to determine the design wind pressure on rooftop solar panels parallel to the roof surface on buildings of all heights and roof slopes.

5.4 Parapets

ASCE/SEI 29.5 refers to ASCE/SEI 27.3.4 and 28.3.2 for wind pressures on parapets for buildings of all heights and low-rise buildings, respectively. See Secs. 3.2.4 and 4.2.2 of this publication for more information on these requirements.

5.5 Roof Overhangs

ASCE/SEI 29.6 refers to ASCE/SEI 27.3.3 and 28.3.3 for wind pressures on roof overhangs for buildings of all heights and for low-rise buildings, respectively. See Secs. 3.2.3 and 4.2.3 of this publication for more information on these requirements.

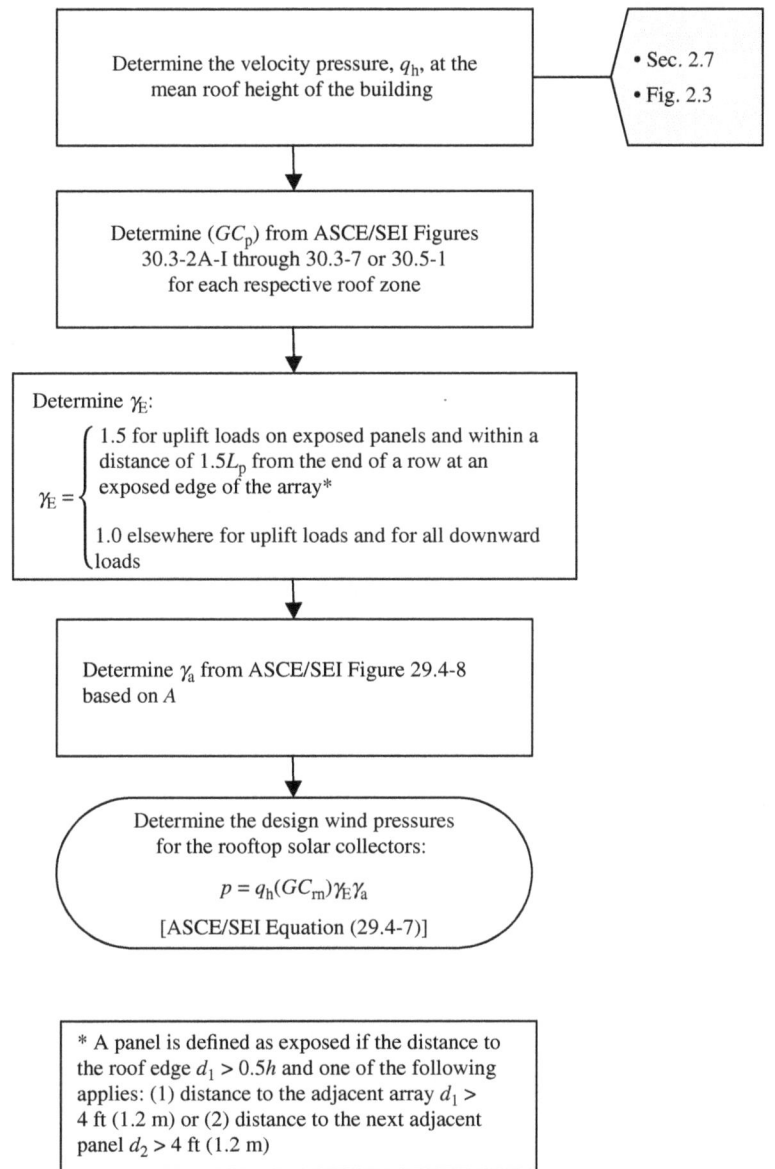

Figure 5.8 Flowchart to determine the design wind pressures on rooftop solar panels parallel to the roof surface on buildings of all heights and roof slopes.

5.6   Minimum Design Wind Loading

The minimum design wind load for other structures must be greater than or equal to 16 lb/ft² (0.77 kN/m²) multiplied by the projected area normal to the wind, A_f (ASCE/SEI 29.7). This load case is to be applied to the structure as a separate load case in addition to the other load cases specified in ASCE/SEI Chapter 29.

5.7 Examples

The following examples illustrate the determination of wind pressures and forces using the Directional Procedure for building appurtenances and other structures in accordance with ASCE/SEI Chapter 29.

5.7.1 Example 5.1—Design Wind Forces on a Solid Freestanding Wall

Determine the design wind forces on a solid freestanding masonry privacy wall that is 60 ft (18.3 m) long and 8 ft (2.4 m) high using the requirements in Chapter 29 and the design data in Table 5.8.

Solution

Check if the freestanding wall meets all the conditions in ASCE/SEI 29.1.2 so that ASCE/SEI Chapter 29 can be used to determine the design wind forces.

The wall is regular-shaped and does not have any unusual geometric irregularities in spatial form. Also, the wall does not have any response characteristics that make it subject to across-wind loading or similar effects, and it is not sited at a location where channeling effects or buffeting in the wake of upwind obstructions need to be considered.

Therefore, the conditions in ASCE/SEI 29.1.2 are met and the requirements in Chapter 29 may be used.

The flowchart in Fig. 5.1 is used to determine the design wind forces.

> *Step 1—Determine the velocity pressure, q_h, at height h*   Fig. 2.3

> The flowchart in Fig. 2.3 is used to determine q_h.

> • Step 1a—Determine the surface roughness category   ASCE/SEI 26.7.2

> In Table 5.8, the surface roughness is given as B.

> • Step 1b—Determine the exposure category   ASCE/SEI 26.7.3

> It is assumed that surface roughness B applies in all directions and that Exposures C and D are not applicable. Therefore, the exposure category is B.

> • Step 1c—Determine the terrain exposure constants   ASCE/SEI Table 26.11-1

> For Exposure B, $\alpha = 7.0$ and $z_g = 1,200$ ft (365.76 m).

> • Step 1d—Determine the velocity pressure exposure coefficient, K_h
>   ASCE/SEI Table 26.10-1

> At $z = 8$ ft (2.4 m), $K_h = 0.57$.

Location	New Haven, CT
Surface roughness	B
Topography	Not situated on a hill, ridge, or escarpment
Occupancy	Low risk to human life in the event of failure
Fundamental natural frequency, n_1	>1 Hz

TABLE 5.8 Design Data for the Solid Freestanding Masonry Privacy Wall in Example 5.1

- Step 1e—Determine the topographic factor, K_{zt} ASCE/SEI 26.8

  Because the wall is not located on a hill, ridge, or escarpment, $K_{zt} = 1.0$.

- Step 1f—Determine the wind directionality factor, K_d
                                                       ASCE/SEI Table 26.6-1

  For a solid freestanding wall, $K_d = 0.85$.

- Step 1g—Determine the ground elevation factor, K_e ASCE/SEI 26.9

  Ground elevation factor can be taken as 1.0 for all elevations.

- Step 1h—Determine the risk category of the wall   ASCE/SEI Table 1.5-1

  This wall represents a low risk to human life in the event of failure, so it falls under Risk Category I.

- Step 1i—Determine the basic wind speed, V Table 2.1

  For Risk Category I, use IBC Figure 1609.3(4) or ASCE/SEI Figure 26.5-1A.

  Equivalently, use Ref. 3 or Ref. 4 to obtain $V = 110$ mi/h (49 m/s) for New Haven, CT.

- Step 1j—Determine the wind velocity pressure, q_h ASCE/SEI 26.10.2

$$q_h = 0.00256 K_h K_{zt} K_d K_e V^2$$

$$= 0.00256 \times 0.57 \times 1.0 \times 0.85 \times 1.0 \times 110^2 = 15.0 \text{ lb/ft}^2$$

In S.I.:

$$q_h = 0.613 K_h K_{zt} K_d K_e V^2$$

$$= 0.613 \times 0.57 \times 1.0 \times 0.85 \times 1.0 \times 49^2 / 1,000 = 0.71 \text{ kN/m}^2$$

Step 2—Determine the gust-effect factor   ASCE/SEI 26.11

The fundamental natural frequency, n_1, of the wall has been determined to be greater than 1 Hz (see Table 5.8). Therefore, the wall is defined as rigid, and the gust-effect factor, G, is permitted to be taken as 0.85.

Step 3—Determine B/s

$$B/s = 60/8 = 7.5$$

In S.I.:

$$B/s = 18.29/2.44 = 7.5$$

Step 4—Determine the gross area of the wall, A_s

$$A_s = Bs = 60 \times 8 = 480 \text{ ft}^2 \ (44.6 \text{ m}^2)$$

Step 5—Determine s/h

For a ground-supported wall, $s/h = 1.0$.

Horizontal Distance from Windward Edge	C_f
0 to s = 8 ft (2.4 m)	3.48
s = 8 ft (2.4 m) to $2s$ = 16 ft (4.9 m)	2.28
$2s$ = 16 ft (4.9 m) to $3s$ = 24 ft (7.3 m)	1.68
$3s$ = 24 ft (7.3 m) to 60 ft (18.3 m)	1.05

TABLE 5.9 Force Coefficients, C_f, for Case C for the Solid Freestanding Wall in Example 5.1

Step 6—Determine ε   ASCE/SEI Figure 29.3-1

Because the wall is solid, $\varepsilon = 1.0$.

Step 7—Determine the force coefficients, C_f, for cases A and B   ASCE/SEI Figure 29.3-1

For $s/h = 1$ and $B/s = 7.5$, $C_f = 1.33$ by linear interpolation.

Step 8—Determine the force coefficients, C_f, for case C   ASCE/SEI Figure 29.3-1

Because $B/s = 7.5 > 2$, case C must be considered.

A summary of the force coefficients over the width of the sign is given in Table 5.9. The values of C_f in the table are obtained by linear interpolation.

Step 9 ——Determine the reduction factor on the force coefficients for case C

According to note 4 in ASCE/SEI Figure 29.3-1, it is permitted to multiply the force coefficients for case C by the reduction factor $1.8 - (s/h) = 1.8 - 1.0 = 0.8$ where $s/h > 0.8$. Reduced force coefficients are given in Table 5.10.

Step 10—Determine the design wind force, F   ASCE/SEI Equation (29.3-1)

For cases A and B:

$$F = q_h G C_f A_s = 15.0 \times 0.85 \times 1.33 \times 480/1,000 = 8.1 \text{ kips}$$

In S.I.:

$$F = 0.71 \times 0.85 \times 1.33 \times 44.6 = 35.8 \text{ kN}$$

For case C:

- From 0 to 8 ft (2.4 m):

$$F = 15.0 \times 0.85 \times 2.78 \times (8.0 \times 8.0)/1,000 = 2.3 \text{ kips}$$

Horizontal Distance from Windward Edge	C_f
0 to s = 8 ft (2.4 m)	2.78
s = 8 ft (2.4 m) to $2s$ = 16 ft (4.9 m)	1.82
$2s$ = 16 ft (4.9 m) to $3s$ = 24 ft (7.3 m)	1.34
$3s$ = 24 ft (7.3 m) to 60 ft (18.3 m)	0.84

TABLE 5.10 Reduced Force Coefficients, C_f, for Case C for the Solid Freestanding Wall in Example 5.1

In S.I.: $F = 0.71 \times 0.85 \times 2.78 \times (2.44 \times 2.44) = 10.0$ kN

- From 8 ft (2.4 m) to 16 ft (4.9 m):

$$F = 15.0 \times 0.85 \times 1.82 \times (8.0 \times 8.0)/1{,}000 = 1.5 \text{ kips}$$

In S.I.:

$$F = 0.71 \times 0.85 \times 1.82 \times (2.44 \times 2.44) = 6.5 \text{ kN}$$

- From 16 ft (4.9 m) to 24 ft (7.3 m):

$$F = 15.0 \times 0.85 \times 1.34 \times (8.0 \times 8.0)/1{,}000 = 1.1 \text{ kips}$$

In S.I.:

$$F = 0.71 \times 0.85 \times 1.34 \times (2.44 \times 2.44) = 4.8 \text{ kN}$$

- From 24 ft (7.3 m) to 60 ft (18.3 m):

$$F = 15.0 \times 0.85 \times 0.84 \times (8.0 \times 36.0)/1{,}000 = 3.1 \text{ kips}$$

In S.I.:

$$F = 0.71 \times 0.85 \times 0.84 \times (2.44 \times 10.97) = 13.6 \text{ kN}$$

Step 11—Apply the design wind forces on the wall   Fig. 5.2

The design wind forces for cases A, B, and C are given in Fig. 5.9.

5.7.2 Example 5.2—Design Wind Forces on a Solid Sign and Support Structure

Determine the design wind forces on the solid sign and support structure depicted in Fig. 5.10 using the requirements in Chapter 29 and the design data in Table 5.11.

Solution

Check if the combined sign and support structure meet all the conditions in ASCE/SEI 29.1.2 so that ASCE/SEI Chapter 29 can be used to determine the design wind forces.

The combined structure is regular-shaped and does not have any unusual geometric irregularities in spatial form. Also, it does not have any response characteristics that make it subject to across-wind loading or similar effects, and it is not sited at a location where channeling effects or buffeting in the wake of upwind obstructions need to be considered.

Therefore, the conditions in ASCE/SEI 29.1.2 are met and the requirements in Chapter 29 may be used.

Part 1: Design Wind Forces on the Solid Sign

The flowchart in Fig. 5.1 is used to determine the design wind forces on the solid sign.

Step 1—Determine the velocity pressure, q_h, at height h   Fig. 2.3

The flowchart in Fig. 2.3 is used to determine q_h.

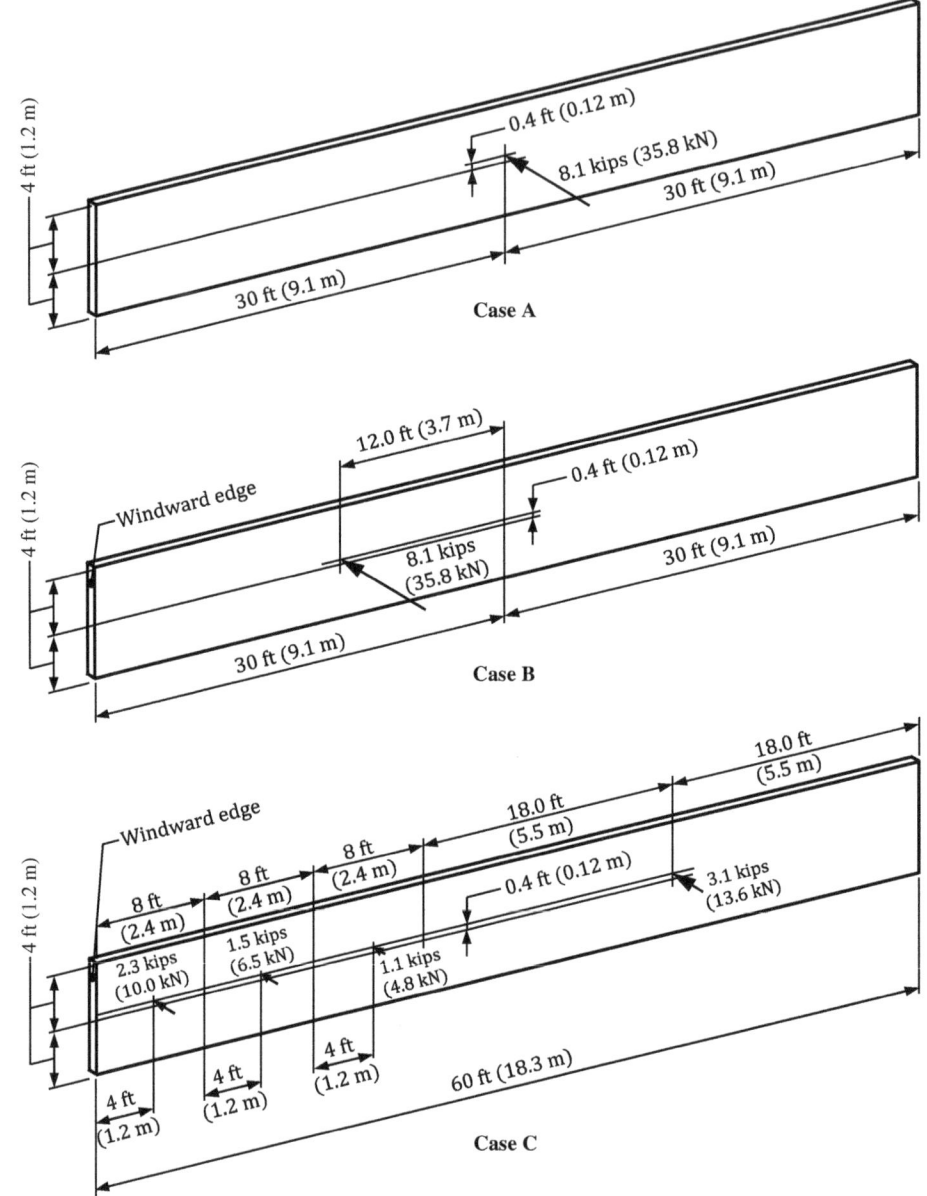

Figure 5.9 Design wind forces, *F*, for the solid freestanding wall in Example 5.1.

- Step 1a—Determine the surface roughness category   ASCE/SEI 26.7.2

  In Table 5.11, the surface roughness is given as C.

- Step 1b—Determine the exposure category   ASCE/SEI 26.7.3

  It is assumed that surface roughness C applies in all directions and that exposures B and D are not applicable. Therefore, the exposure category is C.

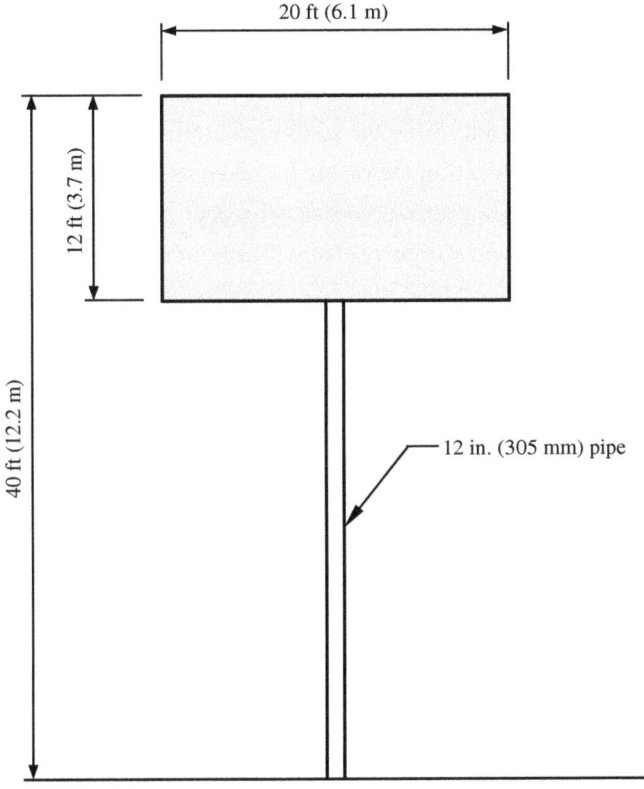

20 ft (6.1 m)

12 ft (3.7 m)

40 ft (12.2 m)

12 in. (305 mm) pipe

FIGURE 5.10 Solid sign and support structure in Example 5.2.

Location	Charlotte, NC
Surface roughness	C
Topography	Not situated on a hill, ridge, or escarpment
Occupancy	Low risk to human life in the event of failure
Fundamental natural frequency, n_1, of the combined structure	>1 Hz

TABLE 5.11 Design Data for the Solid Sign and Support Structure in Example 5.2

- Step 1c—Determine the terrain exposure constants   ASCE/SEI Table 26.11-1
  For Exposure C, $\alpha = 9.5$ and $z_g = 900$ ft (274.32 m).
- Step 1d—Determine the velocity pressure exposure coefficient, K_h
  ASCE/SEI Table 26.10-1

  At $z = 40$ ft (12.2 m), $K_h = 1.04$.
- Step 1e—Determine the topographic factor, K_{zt} ASCE/SEI 26.8
  Because the sign is not located on a hill, ridge, or escarpment, $K_{zt} = 1.0$.

- Step 1f—Determine the wind directionality factor, K_d

    ASCE/SEI Table 26.6-1

    For a solid sign, $K_d = 0.85$.

- Step 1g—Determine the ground elevation factor, K_e ASCE/SEI 26.9

    Ground elevation factor can be taken as 1.0 for all elevations.

- Step 1h—Determine the risk category   ASCE/SEI Table 1.5-1

    This sign and support system represent a low risk to human life in the event of failure, so it falls under Risk Category I.

- Step 1i—Determine the basic wind speed, V Table 2.1

    For Risk Category I, use IBC Figure 1609.3(4) or ASCE/SEI Figure 26.5-1A. Equivalently, use Ref. 3 or Ref. 4 to obtain $V = 103$ mi/h (46 m/s) for Charlotte, NC.

- Step 1j—Determine the wind velocity pressure, q_h ASCE/SEI 26.10.2

$$q_h = 0.00256 K_h K_{zt} K_d K_e V^2$$

$$= 0.00256 \times 1.04 \times 1.0 \times 0.85 \times 1.0 \times 103^2 = 24.0 \text{ lb/ft}^2$$

In S.I.:

$$q_h = 0.613 K_h K_{zt} K_d K_e V^2$$

$$= 0.613 \times 1.04 \times 1.0 \times 0.85 \times 1.0 \times 46^2/1,000 = 1.15 \text{ kN/m}^2$$

Step 2—Determine the gust-effect factor   ASCE/SEI 26.11

The fundamental natural frequency, n_1, of the combined structure has been determined to be greater than 1 Hz (see Table 5.11). Therefore, the structure is defined as rigid, and the gust-effect factor, G, is permitted to be taken as 0.85.

Step 3—Determine B/s

$$B/s = 20/12 = 1.67$$

In S.I.:

$$B/s = 6.10/3.66 = 1.67$$

Step 4—Determine the gross area of the sign, A_s

$$A_s = Bs = 20 \times 12 = 240 \text{ ft}^2 \ (22.3 \text{ m}^2)$$

Step 5—Determine s/h

$$s/h = 12/40 = 0.3$$

In S.I.:

$$s/h = 3.7/12.2 = 0.3$$

Step 6—Determine ε   ASCE/SEI Figure 29.3-1

    Because the sign is solid, $\varepsilon = 1.0$.

Step 7—Determine the force coefficients, C_f, for cases A and B   ASCE/SEI Figure 29.3-1

    For $s/h = 0.3$ and $B/s = 1.67$, $C_f = 1.80$.

Step 8—Determine the force coefficients, C_f, for case C   ASCE/SEI Figure 29.3-1

    Because $B/s = 1.67 < 2$, case C need not be considered.

Step 9—Determine the design wind force, F   ASCE/SEI Equation (29.3-1)

    For cases A and B:

$$F = q_h GC_f A_s = 24.0 \times 0.85 \times 1.80 \times 240/1,000 = 8.8 \text{ kips}$$

In S.I.:

$$F = 1.15 \times 0.85 \times 1.80 \times 22.3 = 39.2 \text{ kN}$$

Step 10—Apply the design wind forces on the sign   Fig. 5.2

    The design wind forces for cases A and B are given in Fig. 5.11.

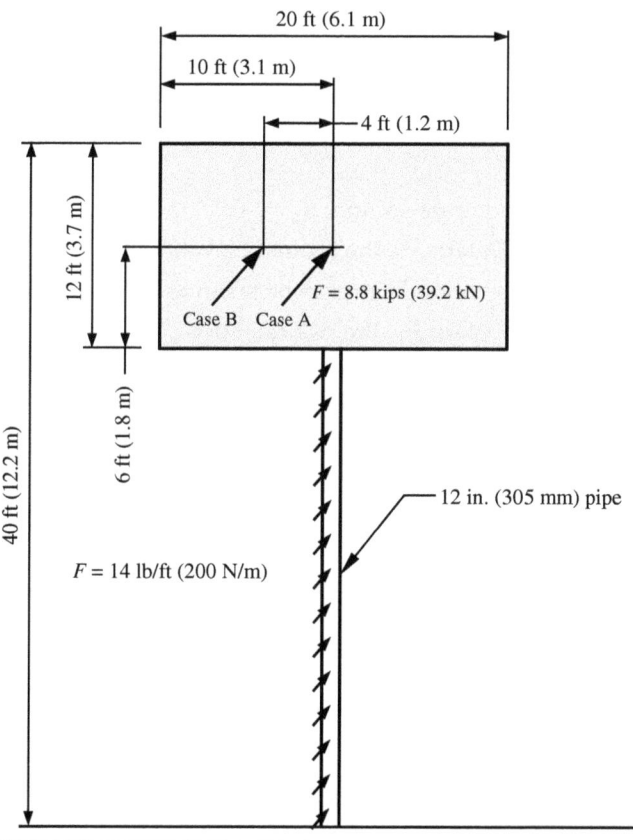

FIGURE 5.11 Design wind forces, *F*, for the solid sign in Example 5.2.

Part 2: Design Wind Forces on the Support Structure

The flowchart in Fig. 5.3 is used to determine the design wind forces on the support structure. It is assumed the force coefficient for the 12-in. (305-mm) diameter pipe is the same as that for a chimney.

> *Step 1—Determine the velocity pressure, q_z, at height z*   Fig. 2.3

> The flowchart in Fig. 2.3 is used to determine q_z.

> - Step 1a—Determine the surface roughness category   ASCE/SEI 26.7.2

> In Table 5.11, the surface roughness is given as C.

> - Step 1b—Determine the exposure category   ASCE/SEI 26.7.3

> It is assumed that surface roughness C applies in all directions and that exposures B and D are not applicable. Therefore, the exposure category is C.

> - Step 1c—Determine the terrain exposure constants
>   ASCE/SEI Table 26.11-1

> For Exposure C, $\alpha = 9.5$ and $z_g = 900$ ft (274.32 m).

> - Step 1d—Determine the velocity pressure exposure coefficient, K_z
>   ASCE/SEI Table 26.10-1

> At $z = 28/2 = 14$ ft (4.3 m), $K_z = 0.85$.

> - Step 1e—Determine the topographic factor, K_{zt} ASCE/SEI 26.8

> Because the structure is not located on a hill, ridge, or escarpment, $K_{zt} = 1.0$.

> - Step 1f—Determine the wind directionality factor, K_d
>   ASCE/SEI Table 26.6-1

> For a round cross-section, $K_d = 1.0$.

> - Step 1g—Determine the ground elevation factor, K_e ASCE/SEI 26.9

> Ground elevation factor can be taken as 1.0 for all elevations.

> - Step 1h—Determine the risk category   ASCE/SEI Table 1.5-1

> This sign and support system represent a low risk to human life in the event of failure, so it falls under Risk Category I.

> - Step 1i—Determine the basic wind speed, V Table 2.1

> For Risk Category I, use IBC Figure 1609.3(4) or ASCE/SEI Figure 26.5-1A. Equivalently, use Ref. 3 or Ref. 4 to obtain $V = 103$ mi/h (46 m/s) for Charlotte, NC.

> - Step 1j—Determine the wind velocity pressure, q_z ASCE/SEI 26.10.2

$$q_z = 0.00256 K_z K_{zt} K_d K_e V^2$$

$$= 0.00256 \times 0.85 \times 1.0 \times 1.0 \times 1.0 \times 103^2 = 23.1 \text{ lb/ft}^2$$

In S.I.:

$$q_z = 0.613 K_z K_{zt} K_d K_e V^2$$

$$= 0.613 \times 0.85 \times 1.0 \times 1.0 \times 1.0 \times 46^2 / 1{,}000 = 1.10 \text{ kN/m}^2$$

Step 2—Determine the gust-effect factor ASCE/SEI 26.11

The fundamental natural frequency, n_1, of the combined structure has been determined to be greater than 1 Hz (see Table 5.11). Therefore, the structure is defined as rigid, and the gust-effect factor, G, is permitted to be taken as 0.85

Step 3—Determine the force coefficient, C_f ASCE/SEI Figure 29.4-1

For a round cross-section, $D\sqrt{q_z} = 1.0 \times \sqrt{23.1} = 4.8 > 2.5$

In S.I.:

$$D\sqrt{q_z} = 0.31 \times \sqrt{1,100} = 10.3 > 5.3$$

$$h/D = 28.0/1.0 = 28.0$$

In S.I.:

$$h/D = 8.53/0.305 = 28.0$$

Therefore, for a moderately smooth surface, $C_f = 0.7$.

Step 4—Determine the design wind force, F ASCE/SEI Equation (29.4-1)

$$F = q_z G C_f A_f = 23.1 \times 0.85 \times 0.7 \times 1.0 = 14 \text{ lb/ft}$$

In S.I.:

$$F = 1.10 \times 1,000 \times 0.85 \times 0.7 \times 0.305 = 200 \text{ N/m}$$

Step 5—Apply the design wind force on the pipe

The design wind force on the pipe is given in Fig. 5.11.

5.7.3 Example 5.3—Design Wind Forces on Rooftop Equipment

Determine the design wind forces on the rooftop equipment depicted in Fig. 5.12 for wind in the N-S direction using the requirements in Chapter 29. The equipment,

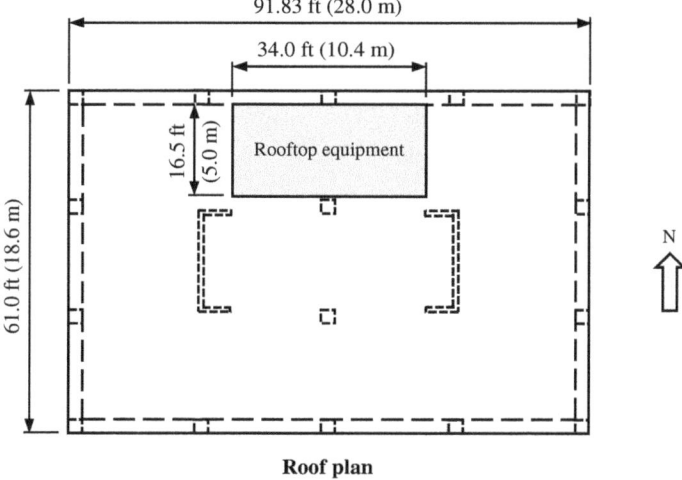

FIGURE 5.12 Rooftop equipment in Example 5.3.

which is 14.0 ft (4.3 m) tall, is located on the roof of the residential building in Example 3.6.

Step 1—Determine the velocity pressure, q_h, at the mean roof height of the building

Fig. 2.3

From Table 3.23 in Example 3.6, $q_h = 27.6$ lb/ft² (1.33 kN/m²).

Step 2—Determine the resultant lateral force, F_h Table 5.2

Vertical projected area of the rooftop unit $A_f = 14.0 \times 34.0 = 476.0$ ft² (44.2 m²)

$$0.1Bh = 0.1 \times 91.83 \times 152.0 = 1,395.8 \text{ ft}^2 \ (129.7 \text{ m}^2) > A_f$$

Therefore, $(GC_r) = 1.9$. ASCE/SEI 29.4.1

$F_h = q_h(GC_r)A_f = 27.6 \times 1.9 \times 476.0/1,000 = 25.0$ kips   ASCE/SEI Equation (29.4-2)

In S.I.:

$$F_h = 1.33 \times 1.9 \times 44.2 = 111.7 \text{ kN}$$

This force is applied at the centroid of A_f (see Fig. 5.4).

Step 3—Determine the vertical uplift force, F_v Table 5.2

Horizontal projected area of the rooftop unit $A_r = 16.5 \times 34.0 = 561.0$ ft² (52.1 m²)

$$0.1BL = 0.1 \times 91.83 \times 61.0 = 560.2 \text{ ft}^2 \ (52.0 \text{ m}^2) < A_r$$

Therefore,

$$(GC_r) = 1.0 + 0.56[1 - (A_r/BL)]$$ Table 5.2

$$= 1.0 + 0.56 \times [1 - (561.0/5,601.6)] = 1.5$$

In S.I.:

$$(GC_r) = 1.0 + 0.56 \times [1 - (52.1/520.4)] = 1.5$$

$F_v = q_h(GC_r)A_r = 27.6 \times 1.5 \times 561.0/1,000 = 23.2$ kips   ASCE/SEI Equation (29.4-2)

In S.I.:

$$F_v = 1.33 \times 1.5 \times 52.1 = 103.9 \text{ kN}$$

This force is applied at the centroid of A_r (see Fig. 5.4).

5.7.4   Example 5.4–Isolated Circular Tank

Determine the design wind forces and pressures on the isolated circular tank depicted in Fig. 5.13 using the requirements in Chapter 29 and the design data in Table 5.12.

Solution

The provisions in ASCE/SEI 29.4.2 are permitted to be used to determine wind forces and pressures on this isolated circular tank because the following conditions are satisfied:

- $h = 25$ ft (7.6 m) < 120 ft (36.6 m)

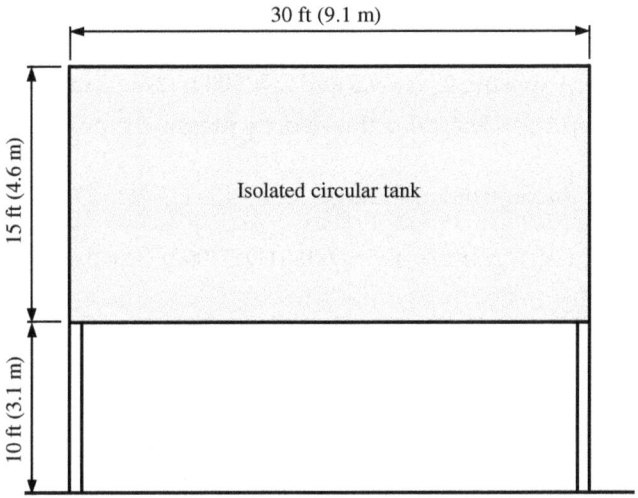

30 ft (9.1 m)

15 ft (4.6 m)

Isolated circular tank

10 ft (3.1 m)

Figure 5.13 Isolated circular tank in Example 5.4.

Location	Tulsa, OK
Surface roughness	C
Topography	Not situated on a hill, ridge, or escarpment
Occupancy	Tank is part of a hospital complex
Fundamental natural frequency, n_1, of the structure	>1 Hz

Table 5.12 Design Data for the Isolated Circular Tank in Example 5.4

- $D = 30$ ft (9.1 m) < 120 ft (36.6 m)
- $0.25 < H/D = 15/30 = 0.5 < 4$ (in S.I.: $0.25 < H/D = 4.6/9.1 = 0.5 < 4$)
- $C = 10$ ft (3.1 m) $< H = 15$ ft (4.6 m)

The flowchart in Fig. 5.6 is used to determine the design wind forces and pressures.

- *Step 1—Determine the velocity pressure, q_z, at height z and at the mean roof height, h*   Fig. 2.3

The flowchart in Fig. 2.3 is used to determine q_z.

- Step 1a—Determine the surface roughness category   ASCE/SEI 26.7.2

In Table 5.12, the surface roughness is given as C.

- Step 1b—Determine the exposure category   ASCE/SEI 26.7.3

It is assumed that surface roughness C applies in all directions and that exposures B and D are not applicable. Therefore, the exposure category is C.

- Step 1c—Determine the terrain exposure constants

  ASCE/SEI Table 26.11-1

  For Exposure C, $\alpha = 9.5$ and $z_g = 900$ ft (274.32 m).
- Step 1d—Determine the velocity pressure exposure coefficient, K_z

  ASCE/SEI Table 26.10-1

  At the centroid of the area A_f, $z = 10 + (15/2) = 17.5$ ft (5.3 m) and

  $$K_z = 2.01(z/z_g)^{2/\alpha} = 2.01 \times (17.5/900)^{2/9.5} = 0.88$$

  In S.I. : $K_z = 2.01 \times (5.3/274.32)^{2/9.5} = 0.88$

  At the mean roof height, $z = 25$ ft (7.6 m) and $K_z = K_h = 0.94$.
- Step 1e—Determine the topographic factor, K_{zt} ASCE/SEI 26.8

  Because the tank is not located on a hill, ridge, or escarpment, $K_{zt} = 1.0$.
- Step 1f—Determine the wind directionality factor, K_d

  ASCE/SEI Table 26.6-1

  For a round tank, $K_d = 1.0$.
- Step 1g—Determine the ground elevation factor, K_e ASCE/SEI 26.9

  Ground elevation factor can be taken as 1.0 for all elevations.

  Step 1h—Determine the risk category   ASCE/SEI Table 1.5-1

  Because this tank is part of hospital complex, it is assigned to Risk Category IV.
- Step 1i—Determine the basic wind speed, V Table 2.1

  For Risk Category IV, use IBC Figure 1609.3(3) or ASCE/SEI Figure 26.5-1D. Equivalently, use Ref. 3 or Ref. 4 to obtain $V = 120$ mi/h (54 m/s) for Tulsa, OK.
- Step 1j—Determine the wind velocity pressures q_z and q_h:

  ASCE/SEI 26.10.2

$$q_z = 0.00256 K_z K_{zt} K_d K_e V^2$$

$$= 0.00256 \times 0.88 \times 1.0 \times 1.0 \times 1.0 \times 120^2 = 32.4 \text{ lb/ft}^2$$

$$q_h = 0.00256 K_h K_{zt} K_d K_e V^2$$

$$= 0.00256 \times 0.94 \times 1.0 \times 1.0 \times 1.0 \times 120^2 = 34.7 \text{ lb/ft}^2$$

In S.I.:

$$q_z = 0.613 K_z K_{zt} K_d K_e V^2$$

$$= 0.613 \times 0.88 \times 1.0 \times 1.0 \times 1.0 \times 54^2 / 1,000 = 1.57 \text{ kN/m}^2$$

$$q_h = 0.613 K_h K_{zt} K_d K_e V^2$$

$$= 0.613 \times 0.94 \times 1.0 \times 1.0 \times 1.0 \times 54^2 / 1,000 = 1.68 \text{ kN/m}^2$$

Step 2—Determine the gust-effect factor   ASCE/SEI 26.11

The fundamental natural frequency, n_1, of the structure has been determined to be greater than 1 Hz (see Table 5.12). Therefore, the structure is defined as rigid, and the gust-effect factor, G, is permitted to be taken as 0.85.

Step 3—Determine the design wind forces on the walls, F   ASCE/SEI Equation (29.4-1)

$$F = 0.63 q_z GDH = 0.63 \times 32.4 \times 0.85 \times 30 \times 15/1,000 = 7.8 \text{ kips}$$

In S.I.:

$$F = 0.63 \times 1.57 \times 0.85 \times 9.1 \times 4.6 = 35.2 \text{ kN}$$

Step 4—Determine the pressure coefficients, C_p, for the roof   ASCE/SEI Figure 29.4-5

For zone 1: $C_p = -0.8$

For zone 2: $C_p = -0.5$

Zone 1 extends from the windward edge of the tank a distance $b = 0.5D = 15$ ft (4.6 m) for $H/D = 0.5$.

Step 5—Determine the internal pressure coefficient, (GC_{pi})   Table 2.9

Assuming the tank is enclosed, $(GC_{pi}) = +0.18, -0.18$.

Step 6—Determine the maximum design wind pressures on the roof, p
                                                        ASCE/SEI Equation (29.4-4)

For Zone 1: $p = q_h[GC_p - (GC_{pi})] = 34.7 \times \{[0.85 \times (-0.8)] - 0.18\} = -29.8 \text{ lb/ft}^2$

For Zone 2: $p = 34.7 \times \{[0.85 \times (-0.5)] - 0.18\} = -21.0 \text{ lb/ft}^2$

In S.I.:

For Zone 1: $p = 1.68 \times \{[0.85 \times (-0.8)] - 0.18\} = -1.45 \text{ kN/m}^2$

For Zone 2: $p = 1.68 \times \{[0.85 \times (-0.5)] - 0.18\} = -1.02 \text{ kN/m}^2$

Step 7—Determine the design wind pressure on the underside of the tank, p

Design wind pressures on the underside of the tank are determined by ASCE/SEI Equation (29.4-4) using $C_p = 0.8$ and -0.6.

For $C_p = 0.8$: $p = q_h[GC_p - (GC_{pi})] = 34.7 \times [(0.85 \times 0.8) + 0.18] = 29.8 \text{ lb/ft}^2$

For $C_p = -0.6$: $p = 34.7 \times \{[0.85 \times (-0.6)] - 0.18\} = -23.9 \text{ lb/ft}^2$

In S.I.:

For $C_p = 0.8$: $p = 1.68 \times [(0.85 \times 0.8) + 0.18] = 1.45 \text{ kN/m}^2$

For $C_p = -0.6$: $p = 1.68 \times \{[0.85 \times (-0.6)] - 0.18\} = -1.16 \text{ lb/ft}^2$

5.7.5   Example 5.5—Rooftop Solar Panels

Determine the design wind pressures on the solar panels depicted in Fig. 5.14 using the requirements in Chapter 29 and the design data in Table 5.13. The solar panels are supported by the roof of a one-story commercial building, which has a 2.5-ft (0.76-m) parapet on all sides.

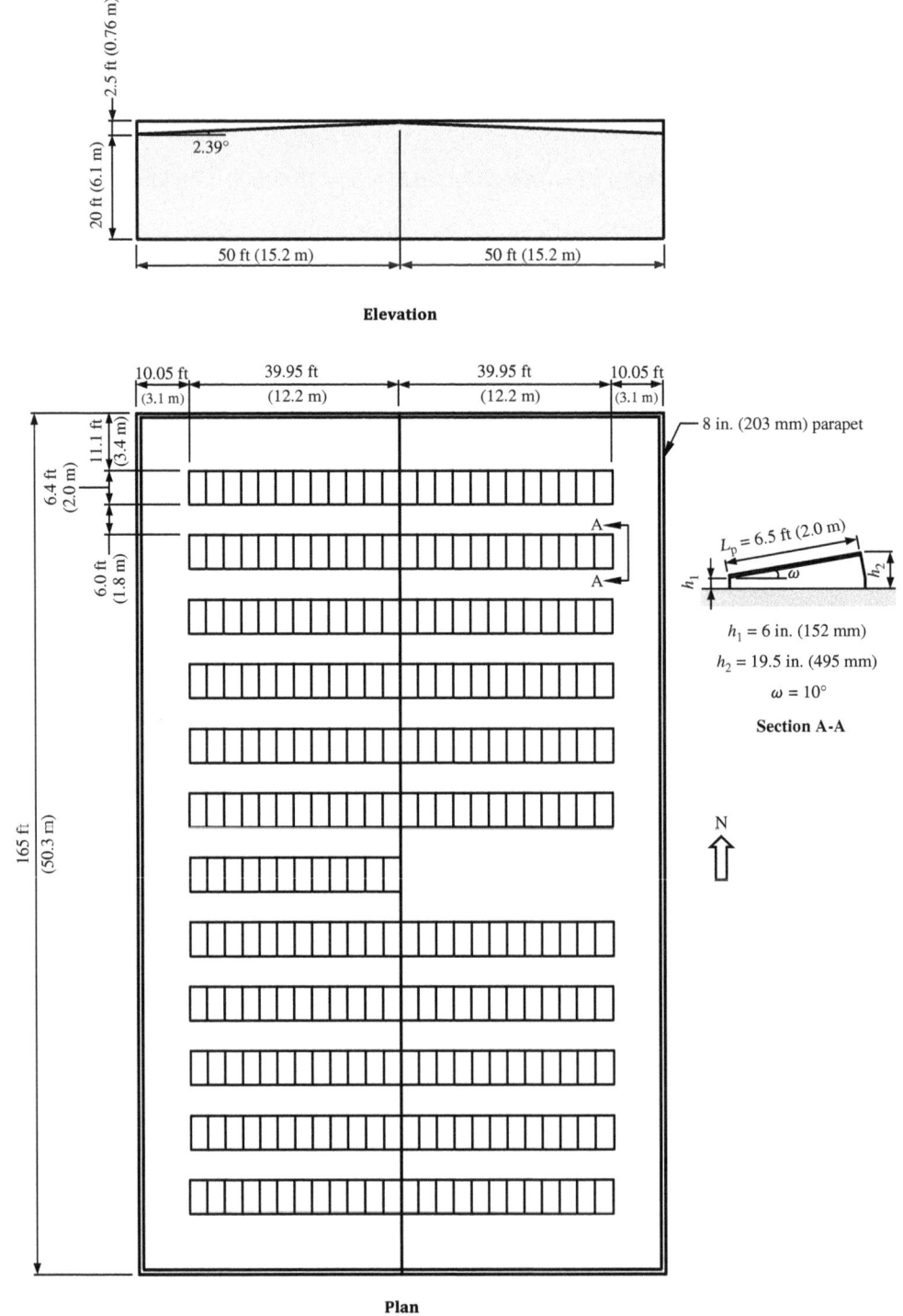

FIGURE. 5.14   Rooftop solar panels in Example 5.5.

Location	Santa Fe, NM
Surface roughness	C
Topography	Not situated on a hill, ridge, or escarpment
Occupancy	Solar panels are supported on the roof of a one-story commercial building
Solar panel dimensions	3.25 ft (1.0 m) by 6.5 ft (2.0 m)
Orientation of solar panels	Portrait
Gap between panels in the E-W direction	1.0 in. (25.4 mm)
Spacing of gaps between panels in the N-S direction	6 ft (1.8 m)

TABLE 5.13 Design Data for the Rooftop Solar Panels in Example 5.5

Solution

Check if the limitations in ASCE/SEI 29.4.3 are satisfied:

- Building is enclosed and has a roof slope = 2.39 degrees < 7 degrees
- Panel chord length $L_p = 6.5$ ft (1.98 m) < 6.7 ft (2.04 m)
- Panel width = 3.25 ft (1.0 m)
- Angle the panel makes with the roof surface $\omega = 10$ degrees < 35 degrees
- Height of the panel above the roof at the lower edge of the panel $h_1 = 0.5$ ft (0.15 m) < 2 ft (0.61 m)
- Height of the panel above the roof at the upper edge of the panel $h_2 = 1.63$ ft (0.50 m) < 4.0 ft (1.2 m)
- Gap between panels (E-W direction) = 1.0 in. (25.4 mm) > 0.25 in. (6.4 mm)
- Spacing of gaps between panels (N-S direction) = 6.0 ft (1.8 m) < 6.7 ft (2.0 m)
- Minimum horizontal clear distance between the panels and the edge of the roof

$$= 10.05 \text{ ft (3.1 m)} > \text{larger of} \begin{cases} 2(h_2 - h_{pt}) = 2 \times (1.63 - 2.5) < 0 \\ \\ 4 \text{ ft (1.2 m)} \end{cases}$$

Therefore, the provisions in ASCE/SEI 29.4.3 are permitted to be used to determine the design wind pressures on the rooftop solar panels.

The flowchart in Fig. 5.7 is used to determine the design wind pressures.

Step 1—Determine the velocity pressure, q_h, at the mean roof height   Fig. 2.3

The flowchart in Fig. 2.3 is used to determine q_h.

- Step 1a—Determine the surface roughness category   ASCE/SEI 26.7.2

  In Table 5.13, the surface roughness is given as C.

- Step 1b—Determine the exposure category   ASCE/SEI 26.7.3

  It is assumed that surface roughness C applies in all directions and that exposures B and D are not applicable. Therefore, the exposure category is C.

- Step 1c—Determine the terrain exposure constants
  ASCE/SEI Table 26.11-1
  For Exposure C, $\alpha = 9.5$ and $z_g = 900$ ft (274.32 m).
- Step 1d—Determine the velocity pressure exposure coefficient, K_h
  ASCE/SEI Table 26.10-1
  At $z = 20$ ft (6.1 m), $K_h = 0.90$.
- Step 1e—Determine the topographic factor, K_{zt} ASCE/SEI 26.8
  Because the building is not located on a hill, ridge, or escarpment, $K_{zt} = 1.0$.
- Step 1f—Determine the wind directionality factor, K_d
  ASCE/SEI Table 26.6-1
  For building structures, $K_d = 0.85$.
- Step 1g—Determine the ground elevation factor, K_e ASCE/SEI 26.9
  Ground elevation factor can be taken as 1.0 for all elevations.
- Step 1h—Determine the risk category   ASCE/SEI Table 1.5-1
  Because of its commercial occupancy, the building is assigned to Risk Category II.
- Step 1i—Determine the basic wind speed, V Table 2.1
  For Risk Category II, use IBC Figure 1609.3(1) or ASCE/SEI Figure 26.5-1B.
  Equivalently, use Ref. 3 or Ref. 4 to obtain $V = 105$ mi/h (47 m/s) for Santa Fe, NM.
- Step 1j—Determine the wind velocity pressure q_h: ASCE/SEI 26.10.2
  $$q_h = 0.00256 K_h K_{zt} K_d K_e V^2$$
  $$= 0.00256 \times 0.90 \times 1.0 \times 0.85 \times 1.0 \times 105^2 = 21.6 \text{ lb/ft}^2$$

In S.I.:
$$q_h = 0.613 \times 0.90 \times 1.0 \times 0.85 \times 1.0 \times 47^2/1{,}000 = 1.04 \text{ kN/m}^2$$

Step 2–Determine the parapet height factor, γ_p   Table 5.5

$$\gamma_p = \text{minimum of} \begin{cases} 1.2 \\ 0.9 + (h_{pt}/h) = 0.9 + (2.5/20) = 1.03 \end{cases}$$

In S.I.:

$$\gamma_p = \text{minimum of} \begin{cases} 1.2 \\ 0.9 + (h_{pt}/h) = 0.9 + (0.76/6.1) = 1.03 \end{cases}$$

Step 3—Determine the panel chord factor, γ_c   Table 5.5

$$\gamma_c = \text{maximum of} \begin{cases} 0.6 + 0.06 L_p = 0.6 + (0.06 \times 6.5) = 0.99 \\ 0.8 \end{cases}$$

Step 4—Determine the exposure factor, γ_E Table 5.5

A panel is defined as exposed where the distance to the roof edge $d_1 > 0.5h = 10$ ft (3.1 m) and one of the following applies:

(1) distance to the adjacent array

$$d_1 > \text{larger of} \begin{cases} 4h_2 = 4 \times 1.63 = 6.5 \text{ ft (2.0 m)} \\ \\ 4 \text{ ft (1.2 m)} \end{cases}$$

(2) distance to the next adjacent panel

$$d_2 > \text{larger of} \begin{cases} 4h_2 = 4 \times 1.63 = 6.5 \text{ ft (2.0 m)} \\ \\ 4 \text{ ft (1.2 m)} \end{cases}$$

Exposed and nonexposed panels are identified in Fig. 5.15.

The exposure factor, γ_E, is equal to the following:

$$\gamma_E = \begin{cases} 1.5 \text{ for uplift loads on exposed panels and within a distance of} \\ 1.5L_p \text{ from the end of a row at an exposed edge of the array} \\ \\ 1.0 \text{ elsewhere for uplift loads and for all downward loads} \end{cases}$$

Step 5—Determine the normalized building length, L_b ASCE/SEI Figure 29.4-7

$$L_b = \text{minimum of} \begin{cases} 0.4(hW_L)^{0.5} = 0.4 \times (20.0 \times 165.0)^{0.5} = 23.0 \text{ ft} \\ \\ h = 20.0 \text{ ft} \\ \\ W_S = 100 \text{ ft} \end{cases}$$

In S.I.:

$$L_b = \text{minimum of} \begin{cases} 0.4(hW_L)^{0.5} = 0.4 \times (6.1 \times 50.3)^{0.5} = 7.0 \text{ m} \\ \\ h = 6.1 \text{ m} \\ \\ W_S = 30.5 \text{ m} \end{cases}$$

Step 6—Determine the normalized wind area, A_n ASCE/SEI Figure 29.4-7

The effective wind area, A, for a solar panel is equal to the area of a panel:

$$A = 3.25 \times 6.5 = 21.1 \text{ ft}^2 \ (1.96 \text{ m}^2)$$

Assuming a panel is supported at its four corners, A for a structural support is equal to $21.1/4 = 5.3$ ft^2 (0.49 m^2).

$$A_n = \frac{1,000A}{\max[L_b, 15 \text{ ft (4.6 m)}]^2}$$

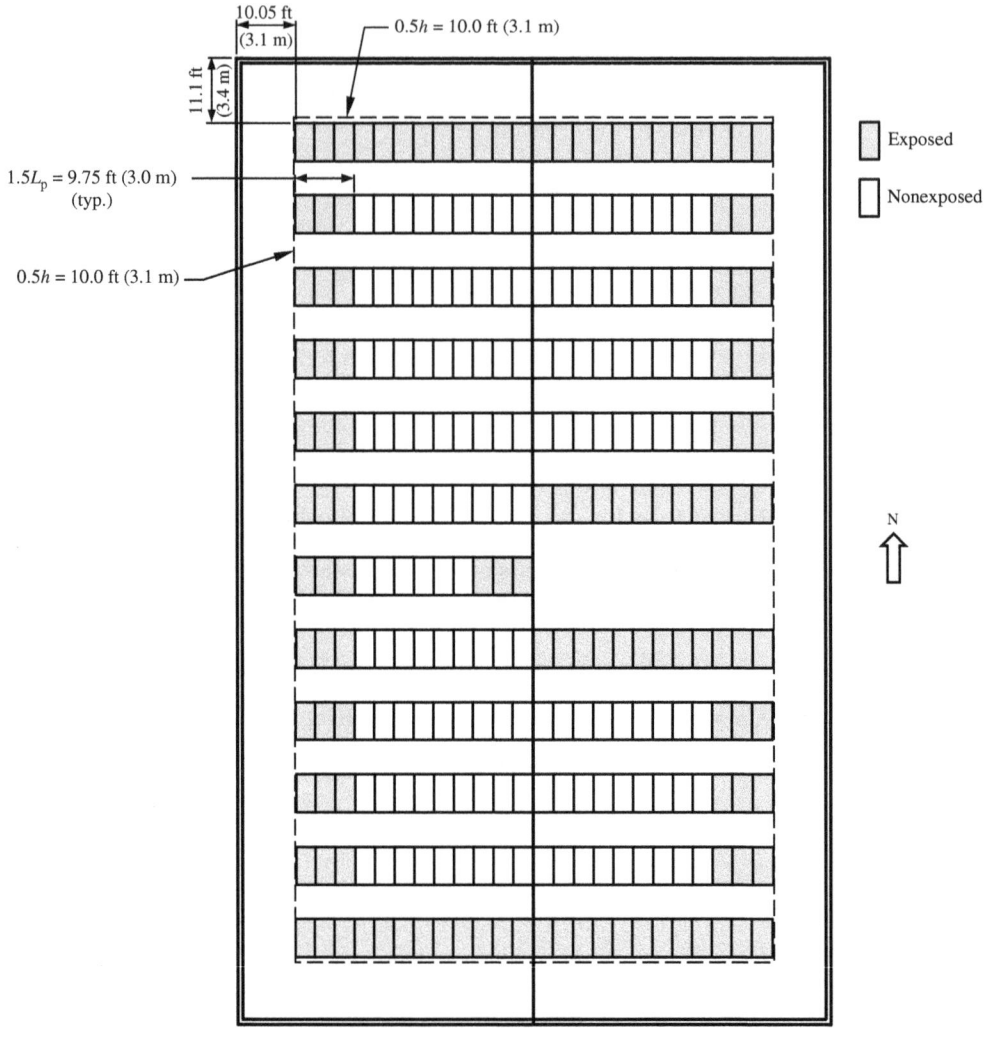

FIGURE 5.15   Exposed and nonexposed panels in Example 5.5.

For a solar panel:

$$A_n = \frac{1,000 \times 21.1}{20.0^2} = 52.8$$

In S.I.:

$$A_n = \frac{1,000 \times 1.96}{6.1^2} = 52.7$$

For a structural support:

$$A_n = \frac{1,000 \times 5.3}{20.0^2} = 13.3$$

In S.I.:

$$A_n = \frac{1,000 \times 0.49}{6.1^2} = 13.2$$

Step 7—Determine the nominal net pressure coefficients, $(GC_m)_{nom}$
ASCE/SEI Figure 29.4-7

Values of $(GC_m)_{nom}$ are determined from ASCE/SEI Figure 29.4-7 for the solar panels and structural supports for zones 1, 2, and 3 identified in the figure (see Fig. 5.16). The widths of zones 2 and 3 in this example are equal to $2h = 40.0$ ft (12.2 m).

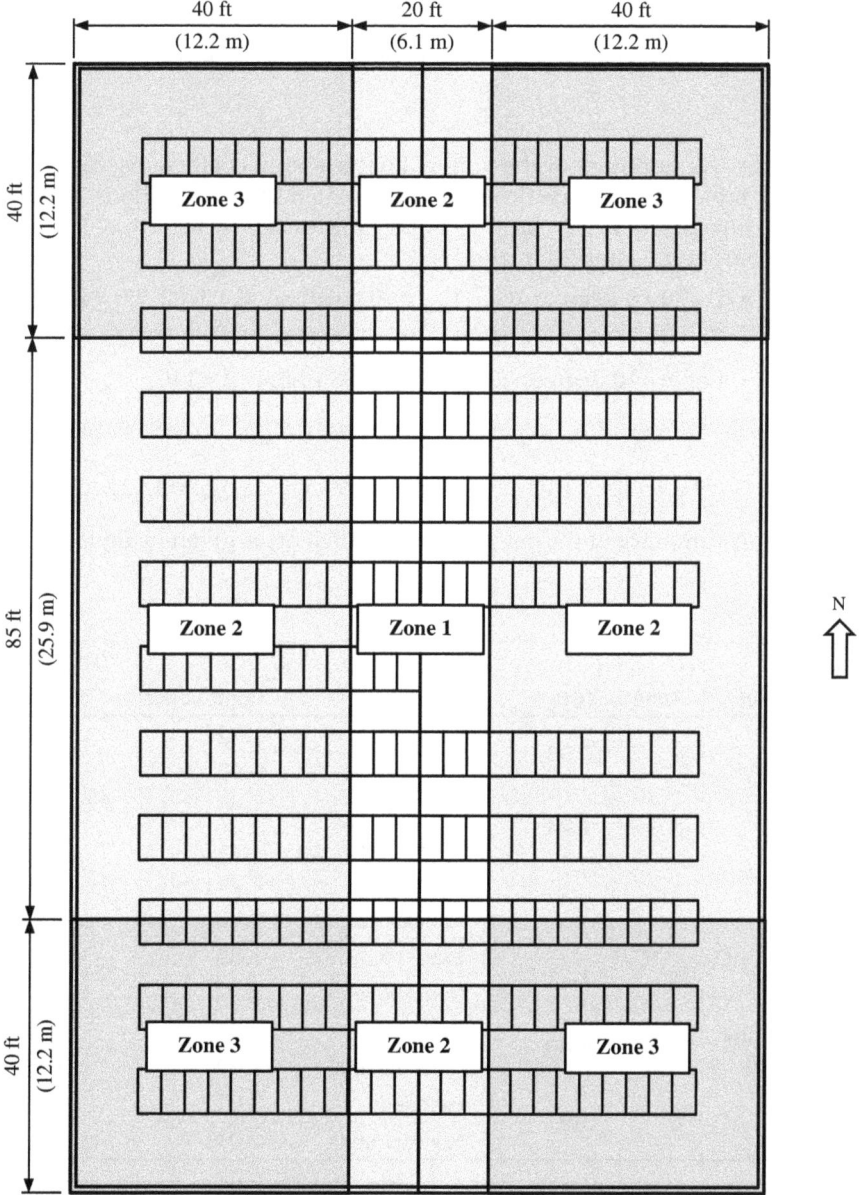

FIGURE 5.16 Roof zones for determining nominal net pressure coefficients, $(GC_{rn})_{nom}$.

Element	A_n	Zone	$(GC_{rn})_{nom}$
Solar panel	52.8	1	0.93
		2	1.24
		3	1.47
Structural support	13.3	1	1.21
		2	1.66
		3	1.97

TABLE 5.14 Nominal Net Pressure Coefficients, $(GC_{rn})_{nom}$, for Example 5.5

A summary of the nominal net pressure coefficients, $(GC_{rn})_{nom}$, is given in Table 5.14. The coefficients are calculated by the equations in Table 5.6. Linear interpolation was used to obtain the values in the table. For example, for a structural support in zone 3:

- For $\omega = 5$ degrees: $(GC_{rn})_{nom} = -0.6669\log_{10}(13.3) + 2.300 = 1.55$
- For $\omega = 15$ degrees: $(GC_{rn})_{nom} = -1.0004\log_{10}(13.3) + 3.500 = 2.38$
- For $\omega = 10$ degrees: $(GC_{rn})_{nom} = (2.38 + 1.55)/2 = 1.97$

Step 8—Determine the net pressure coefficients, (GC_{rn}) ASCE/SEI Equation (29.4-6)

$$(GC_{rn}) = \gamma_p \gamma_c \gamma_E (GC_{rn})_{nom} = 1.03 \times 0.99 \times \gamma_E (GC_{rn})_{nom} = 1.02\gamma_E (GC_{rn})_{nom}$$

A summary of the net pressure coefficients is given in Table 5.15.

Element	Zone	$(GC_{rn})_{nom}$	Panel Type	(GC_{rn}) Uplift Loads	(GC_{rn}) Downward Loads
Solar panel	1	0.93	Exposed	1.42	0.95
			Nonexposed	0.95	0.95
	2	1.24	Exposed	1.90	1.27
			Nonexposed	1.27	1.27
	3	1.47	Exposed	2.25	1.50
			Nonexposed	1.50	1.50
Structural support	1	1.21	Exposed	1.85	1.23
			Nonexposed	1.23	1.23
	2	1.66	Exposed	2.54	1.69
			Nonexposed	1.69	1.69
	3	1.97	Exposed	3.01	2.01
			Nonexposed	2.01	2.01

TABLE 5.15 Net Pressure Coefficients, (GC_{rn}), for Example 5.5

Element	Zone	Panel Type	Uplift Loads		Downward Loads	
			(GC_m)	p, lb/ft² (kN/m²)	(GC_m)	p, lb/ft² (kN/m²)
Solar panel	1	Exposed	1.42	30.7 (1.48)	0.95	20.5 (0.99)
		Nonexposed	0.95	20.5 (0.99)	0.95	20.5 (0.99)
	2	Exposed	1.90	41.0 (1.98)	1.27	27.4 (1.32)
		Nonexposed	1.27	27.4 (1.32)	1.27	27.4 (1.32)
	3	Exposed	2.25	48.6 (2.34)	1.50	32.4 (1.56)
		Nonexposed	1.50	32.4 (1.56)	1.50	32.4 (1.56)
Structural support	1	Exposed	1.85	40.0 (1.92)	1.23	26.6 (1.28)
		Nonexposed	1.23	26.6 (1.28)	1.23	26.6 (1.28)
	2	Exposed	2.54	54.9 (2.64)	1.69	36.5 (1.76)
		Nonexposed	1.69	36.5 (1.76)	1.69	36.5 (1.76)
	3	Exposed	3.01	65.0 (3.13)	2.01	43.4 (2.09)
		Nonexposed	2.01	43.4 (2.09)	2.01	43.4 (2.09)

TABLE 5.16 Design Wind Pressures for the Rooftop Solar Panels and Structural Supports in Example 5.5

Step 9—Determine the design wind pressures, p   ASCE/SEI Equation (29.4-5)

$$p = q_h(GC_m) = 21.6(GC_m) \text{ lb/ft}^2$$

In S.I.:

$$p = 1.04(GC_m) \text{ kN/m}^2$$

A summary of the design wind pressures is given in Table 5.16. These pressures are normal to the surface of the solar panels.

Components and Cladding

6.1 Overview

This chapter contains the requirements in ASCE/SEI Chapter 30 for determining wind pressures on component and cladding (C&C) elements of buildings and structures.

A summary of the wind load procedures in Chapter 30 is given in Table 6.1.

The provisions in Chapter 30 are applicable to buildings that comply with the following (ASCE/SEI 30.1.2):

- The building is regular-shaped, that is, the structure has no unusual geometrical irregularities in spatial form.

- The building does not have response characteristics that make it subject to across-wind loading, vortex shedding, or instability caused by galloping or flutter. Additionally, the building is not located at a site where channeling effects or buffeting in the wake of upwind obstructions warrant special consideration.

Buildings not meeting these conditions must be designed by either recognized literature that documents such wind load effects or by the wind tunnel procedure in ASCE/SEI Chapter 31 (ASCE/SEI 30.1.3).

Reduction in wind pressure due to apparent shielding by surrounding buildings, other structures, or terrain features is not permitted (ASCE/SEI 30.1.4). Such shielding may be modified or completely removed during the lifespan of the building or structure, which could result in significantly higher wind loads.

It is permitted to use the requirements in Chapter 30 to determine design wind loads on air-permeable cladding, including modular vegetative roof assemblies (ASCE/SEI 30.1.5). Loads lower than those determined by Chapter 30 are permitted where demonstrated by approved test data or recognized literature.

6.2 General Requirements

6.2.1 Minimum Design Wind Pressures

The design wind pressure for C&C elements of buildings must not be less than a net pressure of 16 lb/ft^2 (0.77 kN/m^2) acting in either direction normal to the surface (ASCE/SEI 30.2.2). This load case must be considered in addition to the other required load cases in Chapter 30.

Part	Applicability Building/Element Type		Height Limit	Conditions
1	Enclosed, low-rise		$h \le 60$ ft (18.3 m) and $h \le$ least horizontal dimension of the building	• Regular-shaped building • Building does not have response characteristics making it subject to across-wind loading, vortex shedding, instability due to galloping or flutter • Building is not located at a site where channeling effects or buffeting in the wake of upwind obstructions warrant special consideration • Building has a flat, gable, multispan gable, hip, monoslope, stepped, or sawtooth roof
	Partially enclosed, low-rise			
	Enclosed		$h \le 60$ ft (18.3 m)	
	Partially enclosed			
2	Enclosed, low-rise		$h \le 60$ ft (18.3 m) and $h \le$ least horizontal dimension of the building	• Same first three conditions as in Part 1 • Building has a flat, gable, or hip roof
	Enclosed		$h \le 60$ ft (18.3 m)	
3	Enclosed		$h > 60$ ft (18.3 m)	• Same first three conditions as in Part 1 • Building has a flat, pitched, gable, hip, mansard, arched, or domed roof
	Partially enclosed			
4	Enclosed		60 ft (18.3 m) $< h$ ≤ 160 ft (48.8 m)	• Same first three conditions as in Part 1 • Building has a flat, gable, hip, monoslope, or mansard roof
5	Open		Nonc	• Same first three conditions as in Part 1 • Building has a pitched free, monoslope free, or troughed free roof
6	Building appurtenances and rooftop structures and equipment		None	• Same first three conditions as in Part 1 • See ASCE/SEI 30.8 through 30.11 for additional conditions for the various element types
7	Nonbuilding structures	Circular bins, silos, and tanks	$h \le 120$ ft (36.6 m)	• Same first three conditions as in Part 1
		Rooftop solar panels	None	• Same first three conditions as in Part 1 • Buildings with flat roofs or with gable or hip roofs with roof slopes less than or equal to 7 degrees

TABLE **6.1** Wind Load Procedures in ASCE/SEI Chapter 30

6.2.2 Tributary Areas Greater than 700 ft² (65.0 m²)

It is permitted to design C&C elements of buildings with tributary areas greater than 700 ft² (65.0 m²) for wind pressures determined using the provisions for main force resisting systems (MWFRSs) (ASCE/SEI 30.2.3). It is assumed that localized wind effects on elements with tributary areas greater than this limit are not as pronounced as those on elements with smaller tributary areas.

6.2.3 External Pressure Coefficients

Combined gust-effect factor and external pressure coefficients for C&C elements, (GC_p), are given for the various elements covered in Chapter 30. According to ASCE/SEI 30.2.4, these factors and coefficients must not be separated.

6.3 Effective Wind Area

The effective wind area, A, is used throughout Chapter 30 to determine external pressure coefficients, which, in turn, are used to determine design wind pressures on C&C elements.

In accordance with ASCE/SEI 26.2, A is determined as follows for the C&C elements in ASCE/SEI Figures 30.3-1 through 30.3-7, 30.4-1, 30.5-1, and 30.7-1 through 30.7-3:

$$A = \text{larger of} \begin{cases} \text{Area tributary to the element} \\ \\ \text{Span length} \times (\text{Span length}/3) \end{cases} \qquad (6.1)$$

Illustrated in Fig. 6.1 are the effective wind areas for a cladding system (consisting of glazing panels and mullions) and a joist system.

For rooftop solar panels and arrays, A for a structural element is equal to the larger of (1) the tributary area of the element (that is, the span length times the perpendicular distance to the adjacent parallel elements), and (2) the span length of the element times one-third the span length.

The effective wind area for cladding fasteners must not be greater than the area tributary to an individual fastener.

6.4 Low-Rise Buildings (Part 1)

Design wind pressures on C&C elements of low-rise buildings satisfying the conditions in ASCE/SEI 30.1.1(1) (see Table 6.1) and those indicated on the applicable figures are determined by ASCE/SEI Equation (30.3-1):

$$p = q_h[(GC_p) - (GC_{pi})] \qquad (6.2)$$

The velocity pressure, q_h, is determined at the mean roof height of the building, h, in accordance with ASCE/SEI 26.10 (see Sec. 2.7 of this publication). Internal pressure coefficients, (GC_{pi}), are determined in accordance with ASCE/SEI 26.13 (see Sec. 2.10 of this publication). Both positive and negative values of (GC_{pi}) must be considered in order to establish the critical load effects.

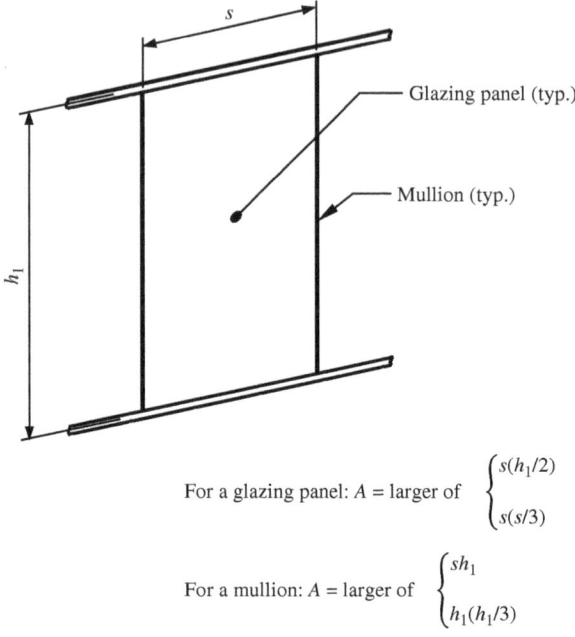

$$\text{For a glazing panel: } A = \text{larger of } \begin{cases} s(h_1/2) \\ s(s/3) \end{cases}$$

$$\text{For a mullion: } A = \text{larger of } \begin{cases} sh_1 \\ h_1(h_1/3) \end{cases}$$

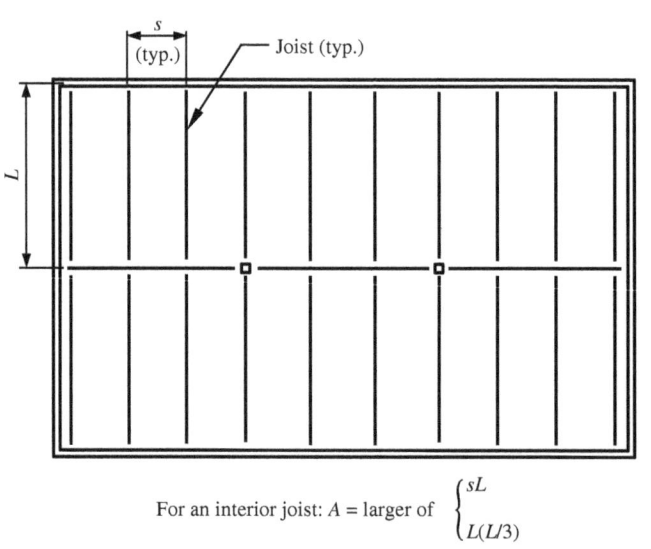

$$\text{For an interior joist: } A = \text{larger of } \begin{cases} sL \\ L(L/3) \end{cases}$$

FIGURE 6.1 Effective wind areas, A, for a cladding system and a joist system.

External pressure coefficients, (GC_p), for C&C elements located on the walls and roof of a building are determined by the ASCE/SEI figures in Table 6.2.

The equations in Tables C30.3-1 through C30.3-10 in ASCE/SEI C30.3.2 can be used to determine the external pressure coefficients, (GC_p), in ASCE/SEI Figures 30.3-1 and 30.3-2A through 30.3-2I.

Element	ASCE/SEI Figure
Walls	30.3-1
Flat, gable, and hip roofs	30.3-2A through 30.3-2I
Stepped roofs	30.3-3
Multispan gable roofs	30.3-4
Monoslope roofs	30.3-5A and 30.3-5B
Sawtooth roofs	30.3-6
Domed roofs	30.3-7
Arched roofs	27.3-3 (Note 4)

TABLE 6.2 External Pressure Coefficients, (GC_p), for Use in Part 1 of Chapter 30

A step-by-step procedure to determine the design wind pressures on C&C elements in accordance with Part 1 in Chapter 30 (ASCE/SEI 30.3.2) is given in Fig. 6.2.

6.5 Low-Rise Buildings, Simplified (Part 2)

Net design wind pressures on C&C elements of low-rise buildings satisfying the conditions in ASCE/SEI 30.1.1(2) [see Table 6.1] are determined by ASCE/SEI Equation (30.4-1):

$$p_{net} = \lambda K_{zt} p_{net30} \qquad (6.3)$$

Design wind pressures, p_{net30}, for C&C elements located in various zones on a building are tabulated in ASCE/SEI Figure 30.4-1 as a function of the basic wind speed, V, and roof angle, θ, for buildings with a mean roof height of 30 ft (9.1 m) located primarily on flat ground in Exposure B. Modifications are made to the tabulated pressures based on actual building height and exposure using the adjustment factor, λ, given in the figure. Tabulated wind pressures must also be modified by the topographic factor, K_{zt}, evaluated at $0.33h$ in accordance with ASCE/SEI 26.8 where applicable.

Shaded areas in ASCE/SEI Figure 30.4-1 indicate that the final wind pressure, including all permitted deductions, must be greater than or equal to the minimum wind pressure of 16 lb/ft² (0.77 kN/m²) in ASCE/SEI 30.2.2.

The flowchart in Fig. 6.3 can be used to determine the design wind pressures on C&C elements in accordance with Part 2 in Chapter 30 (ASCE/SEI 30.4.2).

6.6 Buildings with $h > 60$ ft (18.3 m) (Part 3)

Design wind pressures on C&C elements of buildings satisfying the conditions in ASCE/SEI 30.1.1(3) (see Table 6.1) are determined by ASCE/SEI Equation (30.5-1):

$$p = q(GC_p) - q_i(GC_{pi}) \qquad (6.4)$$

All the terms in this equation, except for the external pressure coefficients, (GC_p), are the same as those for MWFRSs (see Sec. 3.2.1 of this publication). External pressure coefficients, (GC_p), for C&C elements located on the walls and roof of a building are determined in accordance with the ASCE/SEI figures in Table 6.3.

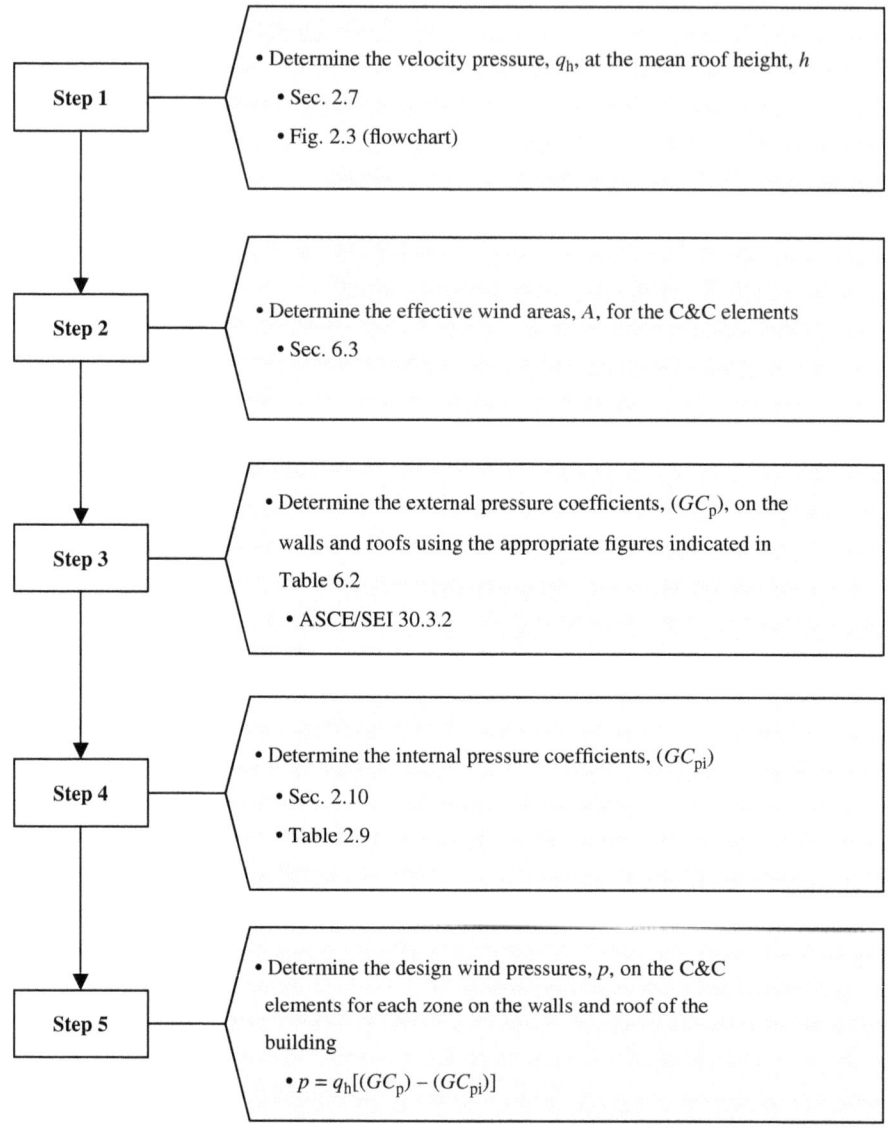

Figure 6.2 Procedure to determine design wind pressures, *p*, in accordance with Part 1 in Chapter 30.

Wind pressures on C&C elements of buildings with a mean roof height greater than 60 ft (18.3 m) and less than 90 ft (27.4 m) are permitted to be determined using the external pressure coefficients from ASCE/SEI Figures 30.3-1 through 30.3-6 in Part 1 of Chapter 30 provided the height-to-width ratio of the building is less than or equal to 1.

A step-by-step procedure to determine the design wind pressures on C&C elements in accordance with Part 3 in Chapter 30 (ASCE/SEI 30.5.2) is given in Fig. 6.4.

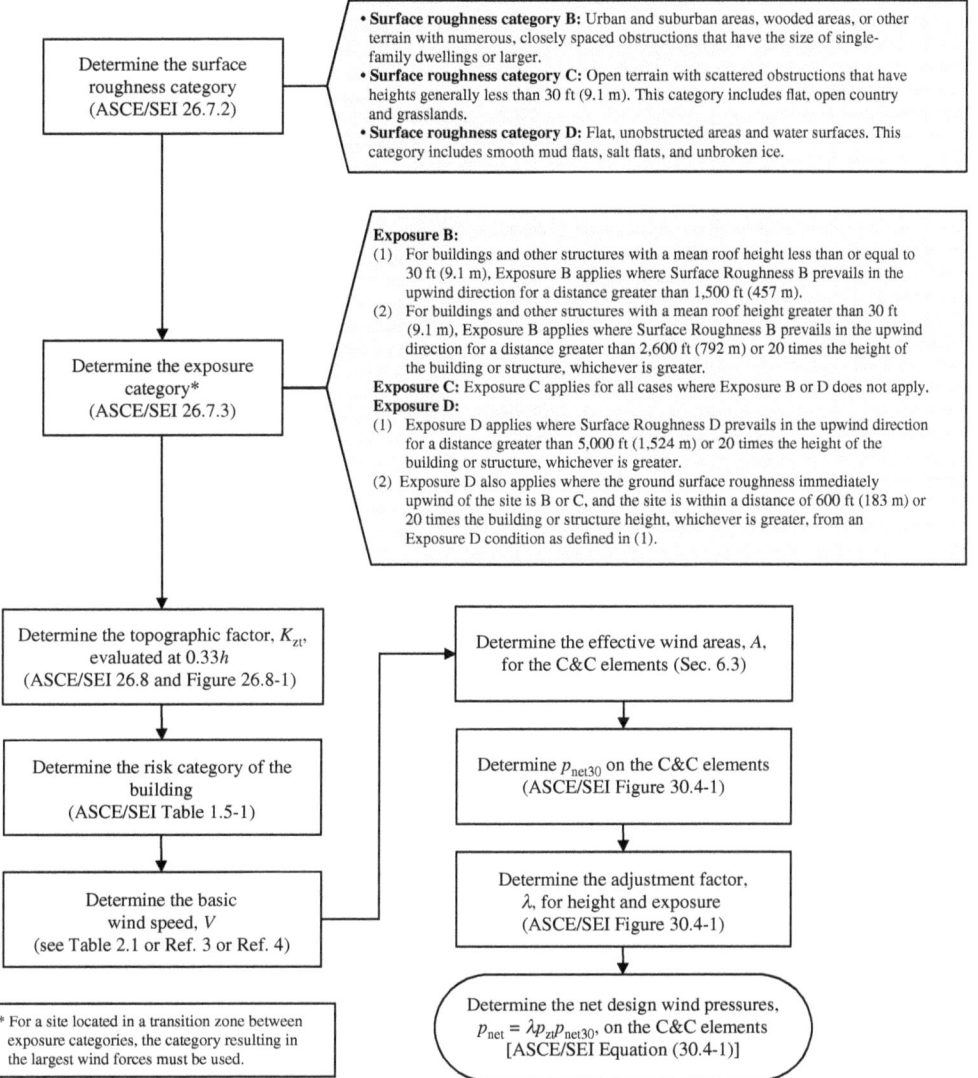

FIGURE 6.3 Flowchart to determine net design wind pressures, p_{net}, in accordance with Part 2 in Chapter 30.

Element	ASCE/SEI Figure
Walls and flat roofs	30.5-1
Arched roofs	27.3-3, Note 4
Domed roofs	30.3-7
Other roof angles and geometries	30.5-1, Note 6

TABLE 6.3 External Pressure Coefficients, (GC_p), for Use in Part 3 of Chapter 30

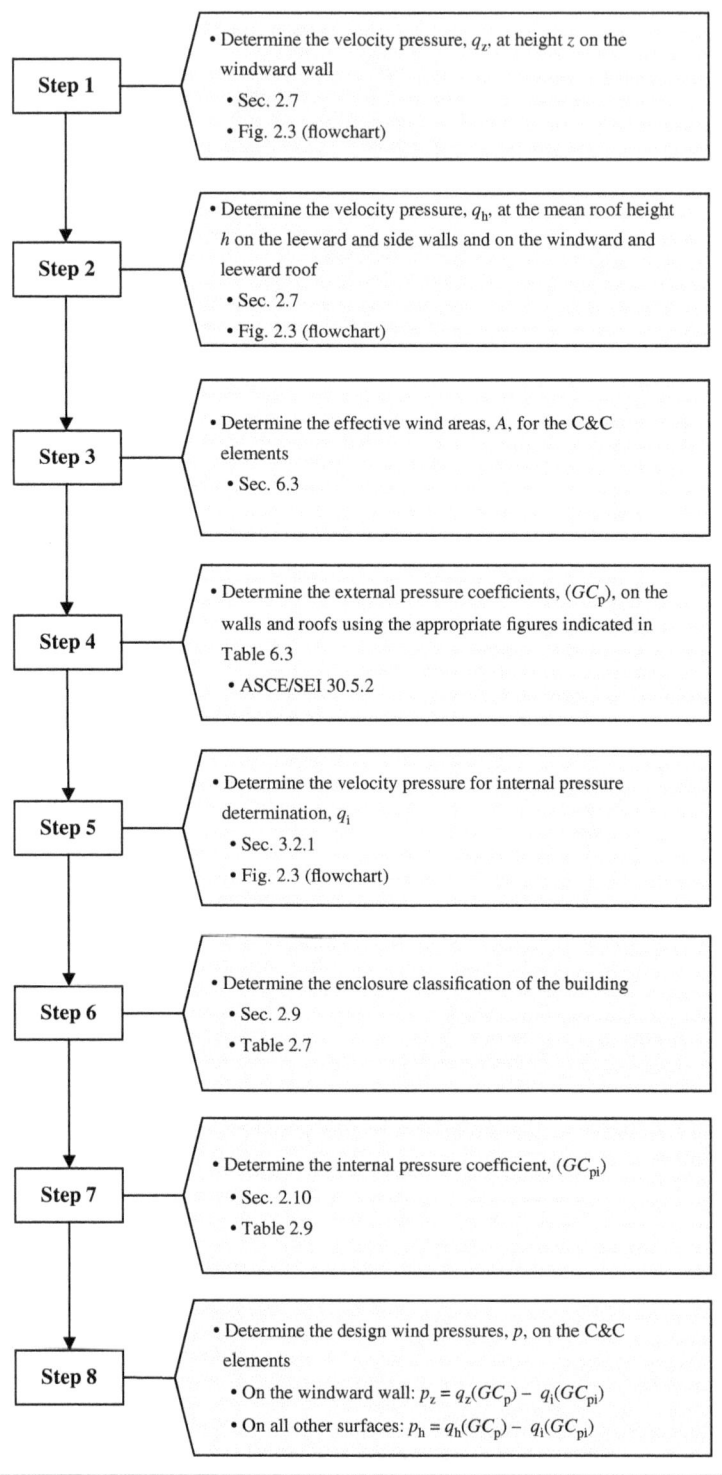

Step 1
- Determine the velocity pressure, q_z, at height z on the windward wall
  - Sec. 2.7
  - Fig. 2.3 (flowchart)

Step 2
- Determine the velocity pressure, q_h, at the mean roof height h on the leeward and side walls and on the windward and leeward roof
  - Sec. 2.7
  - Fig. 2.3 (flowchart)

Step 3
- Determine the effective wind areas, A, for the C&C elements
  - Sec. 6.3

Step 4
- Determine the external pressure coefficients, (GC_p), on the walls and roofs using the appropriate figures indicated in Table 6.3
  - ASCE/SEI 30.5.2

Step 5
- Determine the velocity pressure for internal pressure determination, q_i
  - Sec. 3.2.1
  - Fig. 2.3 (flowchart)

Step 6
- Determine the enclosure classification of the building
  - Sec. 2.9
  - Table 2.7

Step 7
- Determine the internal pressure coefficient, (GC_{pi})
  - Sec. 2.10
  - Table 2.9

Step 8
- Determine the design wind pressures, p, on the C&C elements
  - On the windward wall: $p_z = q_z(GC_p) - q_i(GC_{pi})$
  - On all other surfaces: $p_h = q_h(GC_p) - q_i(GC_{pi})$

Figure 6.4 Procedure to determine design wind pressures, p, in accordance with Part 3 in Chapter 30.

6.7 Buildings with 60 ft (18.3 m) < h ≤ 160 ft (48.8 m), Simplified (Part 4)

6.7.1 Wall and Roof Surfaces

Design wind pressures on C&C elements of buildings satisfying the conditions in ASCE/SEI 30.1.1(4) (see Table 1) are determined by ASCE/SEI Equation (30.6-1):

$$p = p_{table}(EAF)(RF)K_{zt} \qquad (6.5)$$

Wind pressures, p_{table}, on designated zones of wall and roof surfaces are determined from ASCE/SEI Table 30.6-2 based on the basic wind speed, V, mean roof height, h, and roof angle, θ, for buildings located primarily on flat ground and Exposure C. The tabulated pressures are for an effective wind area, A, of 10 ft² (0.93 m²). For buildings with a mean roof height greater than 60 ft (18.3 m) but less than 70 ft (21.3 m), the design wind pressures at 70 ft (21.3 m) must be applied to the wall and roof surfaces (ASCE/SEI 30.6.1.1).

Requirements on the determination of wind pressures for buildings of varying heights, roof forms, and roof angles are also given in ASCE/SEI 30.6.1.1 and are summarized in Table 6.4. In cases where wind pressures are determined using ASCE/SEI Figure 30.4-1, the adjustment factor, λ, must be applied to roof pressures accordingly where λ is given in ASCE/SEI Table 30.6-2.

The exposure adjustment factor (EAF), which is given in ASCE/SEI Table 30.6-2, modifies the tabulated wind pressures in cases where the exposure at the site is different than Exposure C.

The effective area reduction factor (RF), which is also given in ASCE/SEI Table 30.6-2, modifies the tabulated wind pressures for effective wind areas greater than 10 ft² (0.93 m²). Values of RF are provided for designated zones on walls and roof for five different roof shapes and for roof overhangs.

Roof Form	Mean Roof Height, h	Roof Angle, θ	Requirements to Determine Design Wind Pressures
Flat, hip, gable, monoslope, and mansard	≤ 60 ft (18.3 m)	All	Part 2 of Chapter 30 and ASCE/SEI Figure 30.4-1
	> 60 ft (18.3 m)	≤ 7 degrees	ASCE/SEI Table 30.6-2
Hip and gable	> 60 ft (18.3 m)	> 7 degrees	ASCE/SEI Figure 30.4-1 with appropriate q_h (see Note 6 in ASCE/SEI Figure 30.5-1)
Monoslope and mansard	> 60 ft (18.3 m)	> 7 degrees	Refer to "Parameters for Application" in ASCE/SEI Table 30.6-2 for roof zone designations and apply roof pressures from ASCE/SEI Figure 30.4-1 with appropriate q_h (see Note 6 in ASCE/SEI Figure 30.5-1)

TABLE **6.4** Wind Pressure Requirements in ASCE/SEI 30.6.1.1

The topographic factor, K_{zt}, is determined in accordance with ASCE/SEI 26.8 where applicable (see Sec. 2.5 of this publication).

The flowchart in Fig. 6.5 can be used to determine the design wind pressures on C&C wall and roof elements in accordance with Part 4 in Chapter 30 (ASCE/SEI 30.6.1.1).

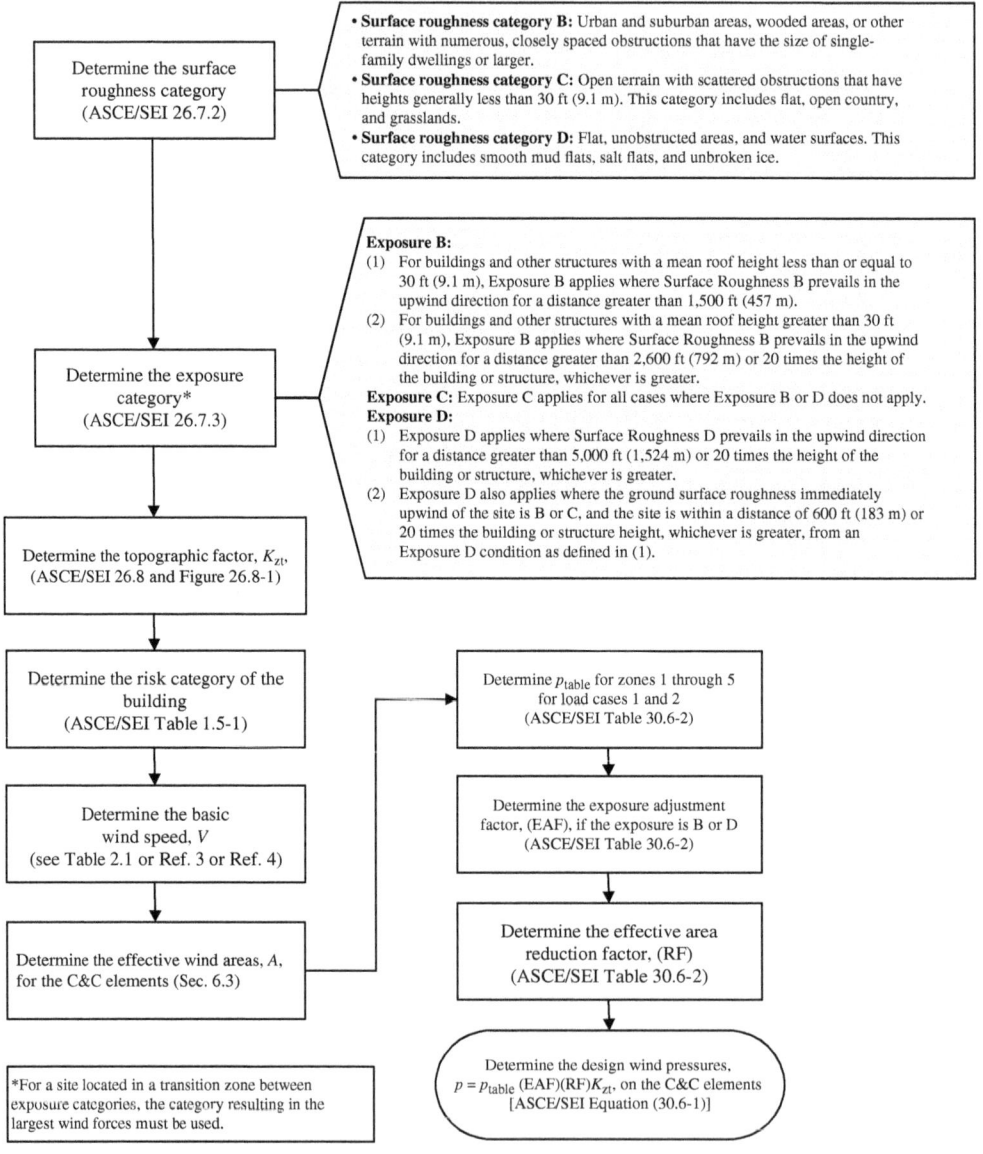

FIGURE 6.5   Flowchart to determine design wind pressures, p, for wall and roof elements in accordance with Part 4 in Chapter 30.

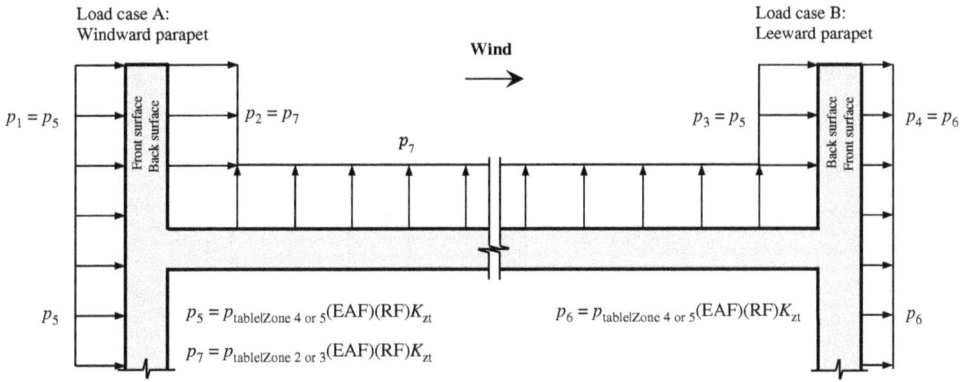

Figure 6.6   Application of wind pressures to parapet C&C elements in accordance with Part 4.

6.7.2   Parapets

ASCE/SEI Equation (30.6-1) is used to determine design wind pressures on parapet C&C elements. The pressures are applied to the parapet in accordance with ASCE/SEI Figure 30.6-1. Two load cases must be considered (see Fig. 6.6):

- Load case A (windward parapets): Windward (front surface) parapet pressure, p_1, is determined using the positive wall pressure, p_5, from zones 4 or 5 from ASCE/SEI Table 30.6-2. Leeward (back surface) parapet pressure, p_2, is determined using the negative roof pressure, p_7, from zones 2 or 3 from ASCE/SEI Table 30.6-2.

- Load case B (leeward parapets): Windward (back surface) parapet pressure, p_3, is determined using the positive wall pressure, p_5, from zones 4 or 5 from ASCE/SEI Table 30.6-2. Leeward (front surface) parapet pressure, p_4, is determined using the negative wall pressure, p_6, from zones 4 or 5 from ASCE/SEI Table 30.6-2.

The height to the top of the parapet must be used in determining p_{table} from ASCE/SEI Table 30.6-2. According to the User Note in ASCE/SEI Figure 30.6-2, C&C roof pressures are permitted to be reduced in accordance with note 5 in ASCE/SEI Figure 30.3-2A and note 7 in ASCE/SEI Figure 30.5-1 when parapets 3 ft (0.91 m) or higher are present.

6.7.3   Roof Overhangs

ASCE/SEI Equation (30.6-1) is used to determine design wind pressures on roof overhang C&C elements. The pressures are applied to the roof overhang in accordance with ASCE/SEI Figure 30.6-2 (see Fig. 6.7).

In zones 1 and 2, the wind pressures on the top surface of the roof overhang are set equal to the pressures calculated by ASCE/SEI Equation (30.6-1). In zone 3, the wind pressures on the top surface are set equal to 1.15 times the pressures calculated by

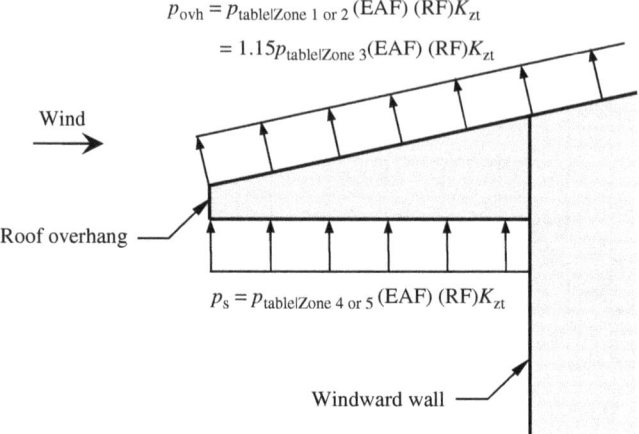

$$P_{\text{ovh}} = P_{\text{table|Zone 1 or 2}} (\text{EAF})(\text{RF})K_{zt}$$

$$= 1.15 P_{\text{table|Zone 3}}(\text{EAF})(\text{RF})K_{zt}$$

Wind

Roof overhang

$$P_s = P_{\text{table|Zone 4 or 5}} (\text{EAF})(\text{RF})K_{zt}$$

Windward wall

FIGURE 6.7 Application of wind pressures to roof overhang C&C elements in accordance with Part 4.

ASCE/SEI Equation (30.6-1). On the underside of the overhang, the wind pressures are set equal to the adjacent wall pressure.

6.8   Open Buildings (Part 5)

Design wind pressures on C&C elements of open buildings satisfying the conditions in ASCE/SEI 30.1.1(5) (see Table 6.1) are determined by ASCE/SEI Equation (30.7-1):

$$p = q_h G C_N \tag{6.6}$$

The velocity pressure, q_h, is determined at the mean roof height of the building, h, in accordance with ASCE/SEI 26.10 (see Sec. 2.7 of this publication). The gust-effect factor, G, is determined in accordance with ASCE/SEI 26.11 (see Sec. 2.8 of this publication).

The figures to be used in determining net pressure coefficients, C_N, for C&C elements on the roofs of open buildings are given in Table 6.5.

A step-by-step procedure to determine the design wind pressures on C&C elements in accordance with Part 5 in Chapter 30 (ASCE/SEI 30.7.2) is given in Fig. 6.8.

Roof Form	ASCE/SEI Figure
Monosloped	30.7-1
Pitched	30.7-2
Troughed	30.7-3

TABLE 6.5 Net Pressure Coefficients for Open Buildings (Part 5)

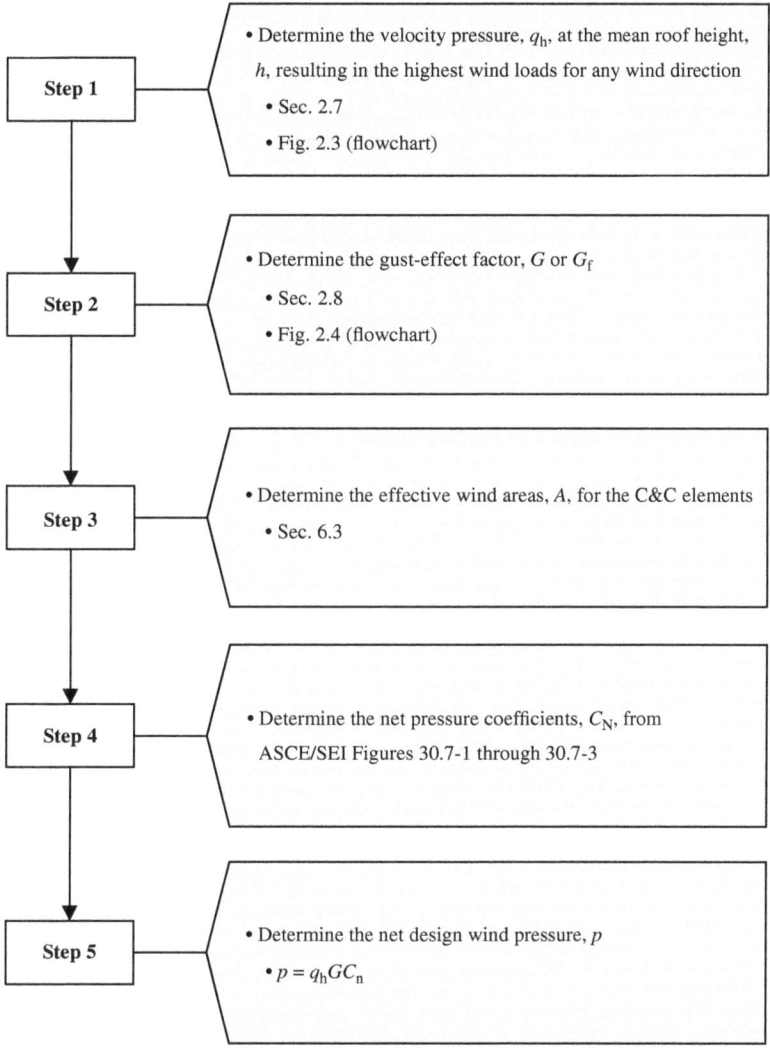

FIGURE 6.8 Procedure to determine design wind pressures, p, in accordance with Part 5 in Chapter 30.

6.9 Building Appurtenances and Rooftop Structures and Equipment (Part 6)

6.9.1 Parapets

Design wind pressures on C&C elements of parapets for all building types and heights, except enclosed buildings with $h \leq 160$ ft (48.8 m) (see Part 4) are determined by ASCE/SEI Equation (30.8-1):

$$p = q_p[(GC_p) - (GC_{pi})] \tag{6.7}$$

In this equation, q_p is the velocity pressure evaluated at the top of the parapet in accordance with ASCE/SEI 26.10 (see Sec. 2.7 of this publications) and (GC_p) is the external pressure coefficient (see Table 6.6).

Internal pressure coefficients, (GC_{pi}), are determined in accordance with ASCE/SEI 26.13 based on the porosity of the parapet envelope (see Sec. 2.10 of this publication). If internal pressure is present, both positive and negative values of (GC_{pi}) must be considered in order to establish the critical load effects.

Load cases A and B must be considered for the windward and leeward parapets, respectively (see ASCE/SEI Figure 30.8-1 and Fig. 6.9):

- Load case A (windward parapets): Windward (front surface) parapet pressure, p_1, is determined using the positive wall pressure, p_5, from zones 4 or 5 in the applicable figure in Table 6.6. Leeward (back surface) parapet pressure, p_2, is determined using the negative roof pressure, p_7, from zones 2 or 3 in the applicable figure in Table 6.6.

C&C Elements	ASCE/SEI Figure
Walls with $h \le 60$ ft (18.3 m)	30.3-1
Flat, gable, and hip roofs	30.3-2A through 30.3-2C
Stepped roofs	30.3-3
Multispan gable roofs	30.3-4
Monoslope roofs	30.3-5A and 30.3-5B
Sawtooth roofs	30.3-6
Domed roofs of all heights	30.3-7
Walls and roofs with $h > 60$ ft (18.3 m)	30.5-1
Arched roofs	27.3-3, note 4

TABLE 6.6 External Pressure Coefficients, (GC_p), for C&C Elements of Parapets

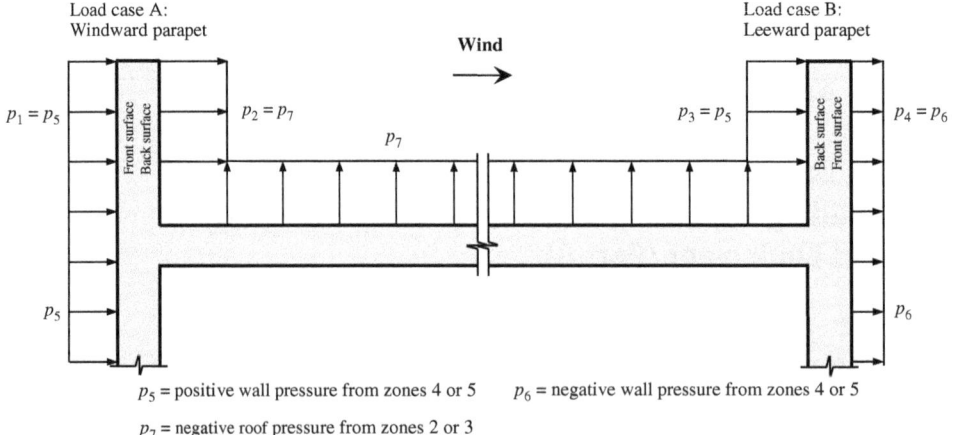

p_5 = positive wall pressure from zones 4 or 5 p_6 = negative wall pressure from zones 4 or 5

p_7 = negative roof pressure from zones 2 or 3

FIGURE 6.9 Application of wind pressures to parapet C&C elements in accordance with Part 6.

- Load case B (leeward parapets): Windward (back surface) parapet pressure, p_3, is determined using the positive wall pressure, p_5, from zones 4 or 5 in the applicable figure in Table 6.6. Leeward (front surface) parapet pressure, p_4, is determined using the negative wall pressure, p_6, from zones 4 or 5 in the applicable figure in Table 6.6.

A step-by-step procedure to determine the design wind pressures on C&C elements of parapets in accordance with Part 6 in Chapter 30 (ASCE/SEI 30.8) is given in Fig. 6.10.

6.9.2 Roof Overhangs

Design wind pressures for roof overhangs of enclosed and partially enclosed buildings of all heights, except enclosed buildings with $h \leq 160$ ft (48.8 m) (see Part 4), are determined by ASCE/SEI Equation (30.9-1):

$$p = q_h[(GC_p) - (GC_{pi})] \tag{6.8}$$

The velocity pressure, q_h, is determined at the mean roof height of the building, h, in accordance with ASCE/SEI 26.10 (see Sec. 2.7 of this publication).

External pressure coefficients, (GC_p), for roof overhangs are given in ASCE/SEI Figures 30.3-2A through 30.3-2C for flat roofs, gable roofs, and hips roofs, respectively.

Internal pressure coefficients, (GC_{pi}), are determined in accordance with ASCE/SEI 26.13.

A step-by-step procedure to determine the design wind pressures on C&C elements of roof overhangs in accordance with Part 6 in Chapter 30 (ASCE/SEI 30.9) is given in Fig. 6.11.

6.9.3 Rooftop Structures and Equipment for Buildings

Design wind pressures on the C&C elements of rooftop structures and equipment for buildings are determined using the requirements in ASCE/SEI 29.4.1, which are applicable in the design of the MWFRS:

- On the walls: Design wind pressure is equal to the horizontal wind load calculated by ASCE/SEI Equation (29.4-2) divided by the vertical projected area of the structure of equipment

- On the roof: Design wind pressure is equal to the vertical wind load calculated by ASCE/SEI Equation (29.4-3) divided by the horizontal projected area of the structure of equipment

The wall pressures can act inward or outward and the roof pressure acts outward.

6.9.4 Attached Canopies on Buildings with $h \leq 60$ ft (18.3 m)

Design wind pressures for canopies attached to the wall of low-rise buildings with $h \leq 60$ ft (18.3 m) are determined by ASCE/SEI Equation (30.11-1):

$$p = q_h(GC_p) \tag{6.9}$$

The velocity pressure, q_h, is determined at the mean roof height of the building, h, in accordance with ASCE/SEI 26.10 (see Sec. 2.7 of this publication).

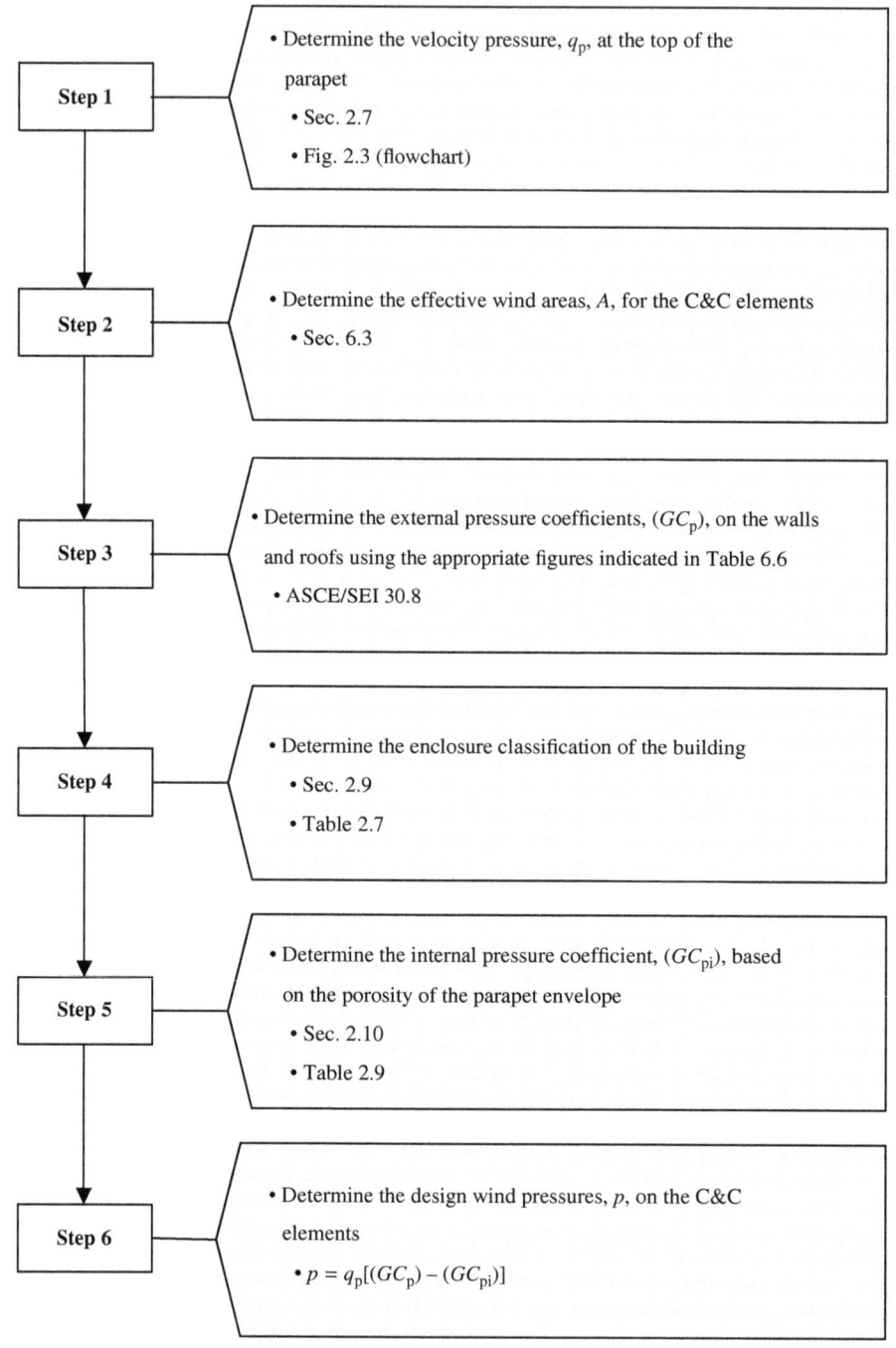

Figure 6.10 Procedure to determine design wind pressures, p, on C&C elements of parapets in accordance with Part 6 in Chapter 30.

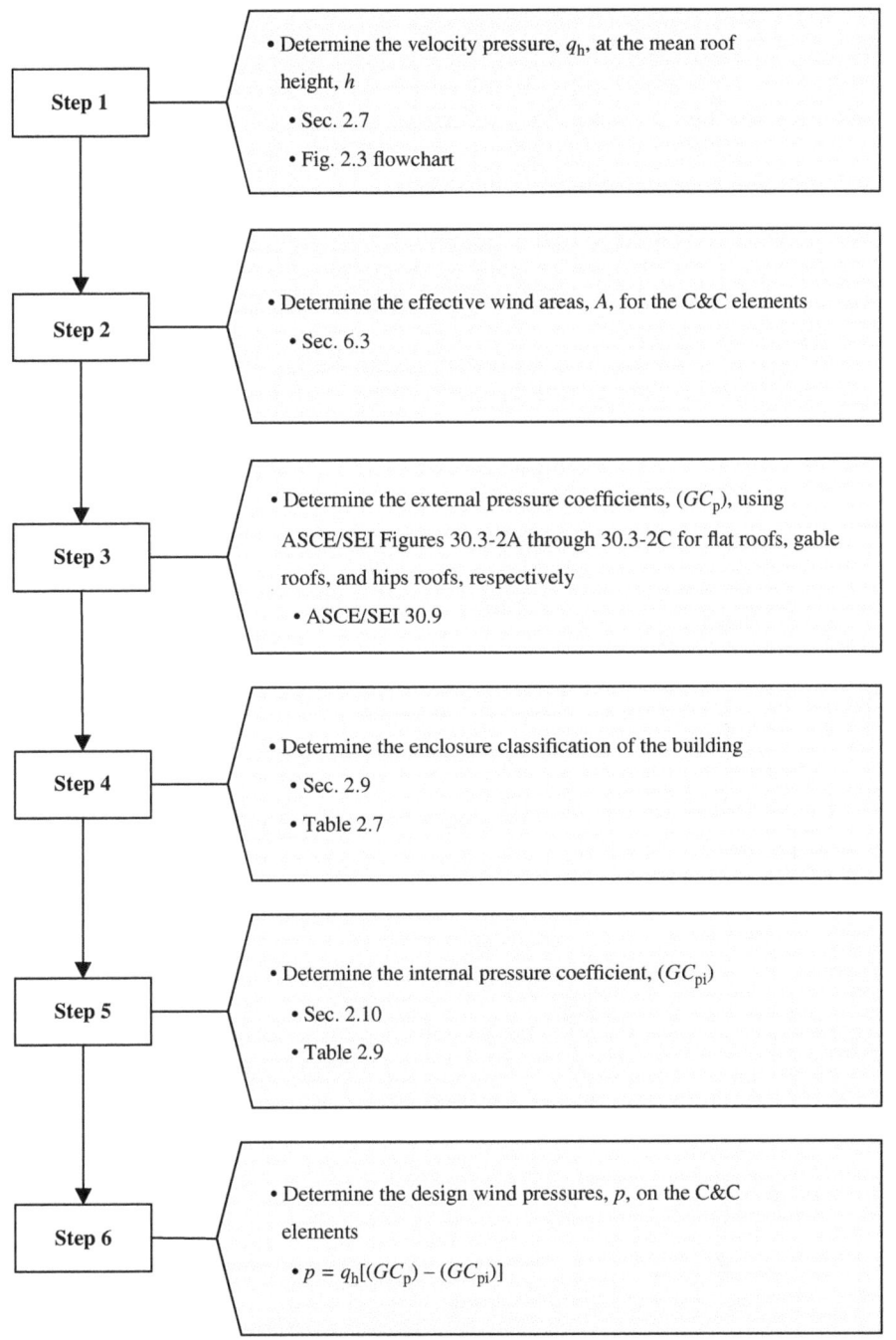

FIGURE 6.11 Procedure to determine design wind pressures, p, on C&C elements of roof overhangs in accordance with Part 6 in Chapter 30.

The net pressure coefficients for attached canopies, (GC_p), are given in ASCE/SEI Figure 30.11-1A where the contributions from the upper and lower surfaces of the canopy are considered individually and in ASCE/SEI Figure 30.11-1B where the contributions of the upper and lower surfaces are considered simultaneously.

ASCE/SEI Figures 30.11-1A and 30.11-1B are both required for canopies with two exposed surfaces. ASCE/SEI Figure 30.11-1A must be used to determine wind pressures on elements and fasteners on the upper and lower surfaces, and ASCE/SEI Figure 30.11-1B is used to determine wind pressures on the structural members of the canopy. ASCE/SEI Figure 30.11-1B is also applicable for canopies with one exposed surface (for example, where wind pressures can be applied directly to the bottom of the top surface only).

A step-by-step procedure to determine the design wind pressures on C&C elements of attached canopies in accordance with Part 6 in Chapter 30 (ASCE/SEI 30.11) is given in Fig. 6.12.

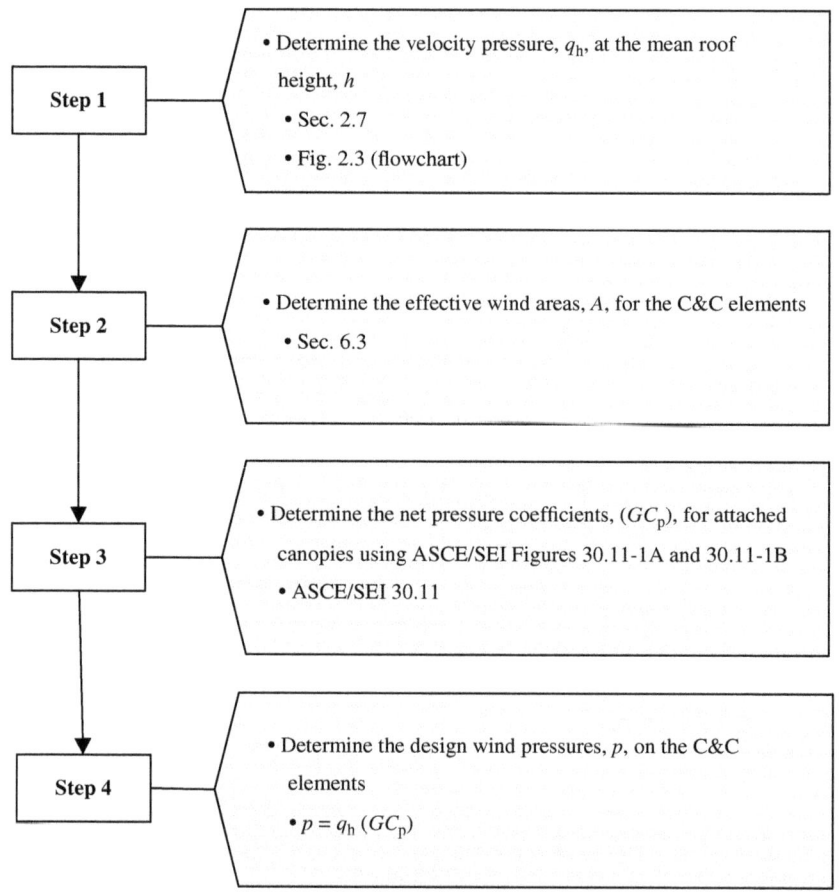

FIGURE 6.12 Procedure to determine design wind pressures, p, on C&C elements of attached canopies in accordance with Part 6 in Chapter 30.

6.10   Nonbuilding Structures (Part 7)

6.10.1   Circular Bins, Silos, and Tanks with $h \leq$ 120 ft (36.6 m)

Design wind pressures on C&C elements for isolated circular bins, silos, and tanks with a mean roof height less than or equal to 120 ft (36.6 m) are determined by ASCE/SEI Equation (30.12-1):

$$p = q_h[(GC_p) - (GC_{pi})] \tag{6.10}$$

The velocity pressure, q_h, is determined at the mean roof height of the structure, h, in accordance with ASCE/SEI 26.10 (see Sec. 2.7 of this publication).

External pressure coefficients, (GC_p), are given in ASCE/SEI 30.12.2 for walls, ASCE/SEI 30.12.5 for underneath sides of column-supported structures (where applicable), and ASCE/SEI 30.12.4 for roofs of isolated circular bins, silos, and tanks. Internal pressure coefficients, (GC_{pi}), are given in ASCE/SEI Table 26.13-1 and ASCE 30.12.3. Required (GC_p) and (GC_{pi}) are given in Table 6.7.

Structures are assumed to be grouped where the center-to-center spacing is less than 1.25 times the diameter of the structure, D. In such cases, the external pressure coefficients for the roof and walls of the grouped structures are given in ASCE/SEI 30.12.6 (see Table 6.8).

The flowchart in Fig. 6.13 can be used to determine the design wind pressures on C&C elements of circular bins, silos, and tanks in accordance with Part 7 in Chapter 30 (ASCE/SEI 30.12).

6.10.2   Rooftop Solar Panels for Buildings of All Heights with Flat Roofs or Gable and Hip Roofs with Slopes Less Than 7 Degrees

The design wind pressures on the C&C elements of rooftop solar modules and panels conforming to the geometric requirements of ASCE/SEI 29.4.3 are determined by the requirements in that section, which are applicable to the MWFRS (see ASCE/SEI 30.13 and Sec. 5.3.4 of this publication).

6.11   Examples

The following examples illustrate the determination of wind pressures on C&C elements in accordance with ASCE/SEI Chapter 30.

6.11.1   Example 6.1—Design Wind Pressures on C&C Elements of a Commercial Building, Chapter 30, Part 1

Determine the design wind pressures on (1) the 7-in. (178-mm) precast concrete wall and (2) the roof joists for the commercial building in Example 3.1 (see Fig. 3.13) using Part 1 in Chapter 30 and the design data Table 3.2. The roof joists are spaced 8 ft (2.4 m) on center.

Element	Pressure Coefficients
External walls	External pressure coefficients: $(GC_{p(\alpha)}) = k_b C_{(\alpha)}$ * where: $k_b = \begin{cases} 1.0 \text{ for } C_{(\alpha)} \geq -0.15 \\\\ 1.0 - 0.55[C_{(\alpha)} + 0.15]\log_{10}(H/D) \text{ for } C_{(\alpha)} < -0.15 \end{cases}$ $C_{(\alpha)} = -0.5 + 0.4\cos\alpha + 0.8\cos 2\alpha + 0.3\cos 3\alpha - 0.1\cos 4\alpha$ $\quad\quad - 0.05\cos 5\alpha$ α = angle from the wind direction to the point on a wall of a circular bin, silo, or tank in degrees
Open-topped circular bins, silos, and tanks	Internal surface of external walls: $(GC_{pi}) = -0.9 - 0.35\log_{10}(H/D)$
Circular bins, silos, and tanks that are not open-topped	Internal pressure coefficients determined in accordance with ASCE/SEI Table 26.13-1
Roofs or lids	External pressure coefficients, (GC_p), determined by ASCE/SEI Figure 30.12-2 for the following (see ASCE/SEI Figure C30.12-2): • Class 1: Flat roofs and conical roofs where the roof angle is less than 10 degrees, and for domed roofs where the average roof angle is less than 10 degrees • Class 2a: Conical roofs where the roof angle is between 10 and 15 degrees inclusive • Class 2b: Conical roofs where the roof angle is greater than 15 degrees and less than or equal to 30 degrees
Undersides of circular bins, silos, and tanks	External pressure coefficients, (GC_p) [see ASCE/SEI Figure 30.12-2]: $(GC_p) = \begin{cases} 1.2 \text{ and } -0.9 \text{ for zone 3} \\\\ 0.8 \text{ and } -0.6 \text{ for zones 1 and 2} \end{cases}$

*Applicable to circular bins, silos, and tanks standing on the ground or supported by columns where $C < H$ (see ASCE/Figure 30.12-1) and $0.25 \leq H/D \leq 4.0$.

TABLE 6.7 Summary of External and Internal Pressure Coefficients for Isolated Circular Bins, Silos, and Tanks

Element	External Pressure Coefficients, (GC_p)
Roofs	ASCE/SEI Figure 30.12-3 for zones 1, 2, 3a, 3b, and 4
Walls	ASCE/SEI Figure 30.12-4 for zones 5a, 5b, 8, and 9

TABLE 6.8 Summary of External Pressure Coefficients for Grouped Circular Bins, Silos, and Tanks

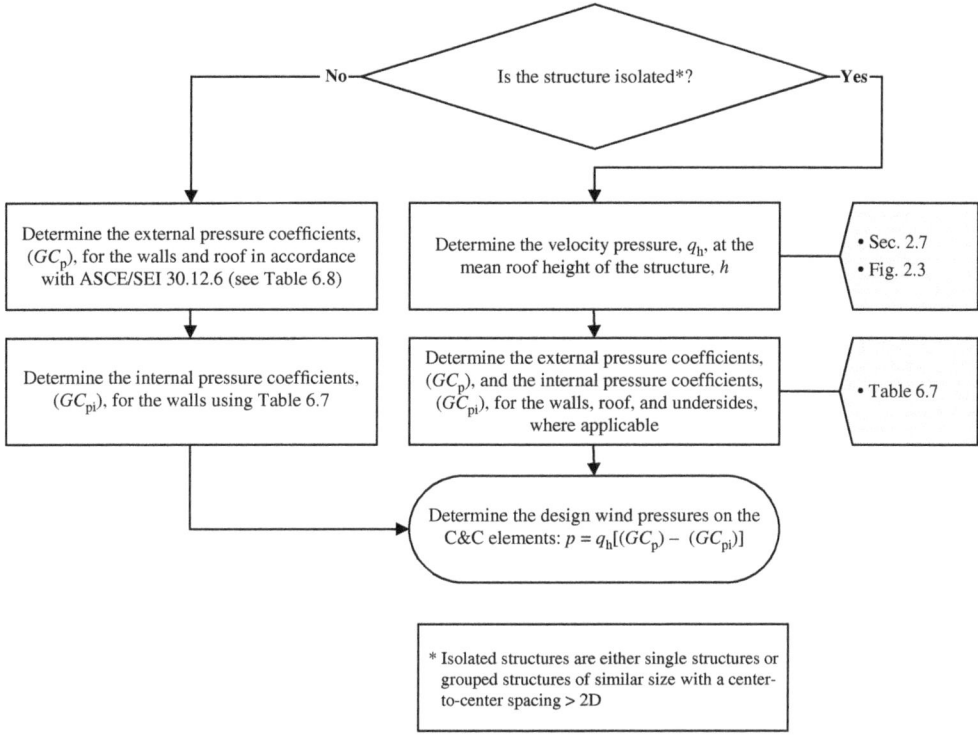

FIGURE 6.13 Flowchart to determine design wind pressures, p, on C&C elements of circular bins, silos, and tanks in accordance with Part 7 in Chapter 30.

Solution

Check if the building meets all the conditions in ASCE/SEI 30.1.1, 30.1.2, and 30.3.1 so that Part 1 in Chapter 30 can be used to determine the design wind pressures on the C&C elements:

- The building is enclosed (see Table 3.2).
- The building is a low-rise building because $h \leq 60$ ft (18.3 m) and is less than the minimum horizontal dimension.
- The building has a flat roof.
- The building is regular-shaped, that is, the structure has no unusual geometrical irregularities in spatial form.
- The building does not have response characteristics that make it subject to across-wind loading, vortex shedding, or instability caused by galloping or flutter. Additionally, the building is not located at a site where channeling effects or buffeting in the wake of upwind obstructions warrant special consideration.

Therefore, the conditions in ASCE/SEI 30.1.1, 30.1.2, and 30.3.1 are met and the requirements in Part 1 of Chapter 30 may be used.

The design procedure in Fig. 6.2 is used to determine the design wind pressures, p, on the C&C elements.

Part 1—Determine the design wind pressures on the 7-in. (178-mm) precast concrete wall

Step 1—Determine the velocity pressure, q_h, at the mean roof height   Fig. 2.3

From Example 3.1, $q_h = 22.6$ lb/ft^2 (1.08 kN/m^2) (see Table 3.4).

Step 2—Determine the effective wind area, A   Sec. 6.3

The effective wind area for the precast concrete wall, which is supported at the ground and at the floor and roof diaphragms, is determined as follows:

$$A = \text{span} \times (\text{span}/3) = 12 \times (12/3) = 48 \text{ ft}^2 \ (4.5 \text{ m}^2)$$

Step 3—Determine the external pressure coefficients, (GC_p)   ASCE/SEI Figure 30.3-1

The external pressure coefficients on the wall in zones 4 and 5 are given in Table 6.9. The equations in ASCE/SEI Table C30.3-1 can be used to determine the external pressure coefficients.

According to note 5 in ASCE/SEI Figure 30.3-1, values of (GC_p) for walls are permitted to be reduced by 10 percent when the roof angle, θ, is less than or equal to 10 degrees. Modified values of (GC_p) based on note 5 are given in Table 6.10.

The width of zone 5, a, is determined as follows:

$$a = \text{smaller of} \begin{cases} 0.1 \times \text{least horizontal dimension} = 0.1 \times 50 = 5 \text{ ft (1.5 m)} \\ \\ 0.4h = 0.4 \times 36 = 14.4 \text{ ft (4.4 m)} \end{cases}$$

$$\text{Minimum } a = \text{greater of} \begin{cases} 0.04 \times \text{least horizontal dimension} = 0.04 \times 50 = 2 \text{ ft (0.61 m)} \\ \\ 3 \text{ ft (0.91 m)} \end{cases}$$

Therefore, $a = 5$ ft (1.5 m).

Zone	(GC$_p$)	
	Positive	Negative
4	0.88	−0.98
5	0.88	−1.16

TABLE 6.9 External Pressure Coefficients, (GC_p), for the Walls in Example 6.1

Zone	(GC$_p$)	
	Positive	Negative
4	0.79	−0.88
5	0.79	−1.04

TABLE 6.10 Modified External Pressure Coefficients, (GC_p), for the Walls in Example 6.1

Step 4—Determine the internal pressure coefficients, (GC_{pi}) Table 2.9

From Step 7 in Example 3.1, $(GC_{pi}) = +0.18, -0.18$ for an enclosed building.

Step 5—Determine the design wind pressures, p   ASCE/SEI Equation (30.3-1)

$$p = q_h[(GC_p) - (GC_{pi})]$$

The maximum design wind pressures on the precast concrete wall are given in Table 6.11. These pressures act perpendicular to the face of the walls. Design wind pressures for zone 4 are as follows:

- For positive (GC_p): $p = 22.6 \times [0.79 - (-0.18)] = 21.9 \text{ lb/ft}^2$
- For negative (GC_p): $p = 22.6 \times [(-0.88) - (+0.18)] = -24.0 \text{ lb/ft}^2$

In S.I.:

- For positive (GC_p): $p = 1.08 \times [0.79 - (-0.18)] = 1.05 \text{ kN/m}^2$
- For negative (GC_p): $p = 1.08 \times [(-0.88) - (+0.18)] = -1.15 \text{ kN/m}^2$

The maximum design wind pressures in both zones are greater than the minimum pressure of 16 lb/ft² (0.77 kN/m²) acting in either direction normal to the surface (ASCE/SEI 30.2.2).

Part 2—Determine the design wind pressures on the roof joists

Step 1—Determine the velocity pressure, q_h, *at the mean roof height*   Fig. 2.3

From Example 3.1, $q_h = 22.6 \text{ lb/ft}^2$ (1.08 kN/m²) (see Table 3.4).

Step 2—Determine the effective wind area, A   Sec. 6.3

The effective wind area for a typical interior roof joist spaced 8 ft (2.4 m) on center is determined as follows (see Fig. 3.13):

$$A = \text{larger of} \begin{cases} 25 \times 8 = 200 \text{ ft}^2 \ (18.6 \text{ m}^2) \\ \\ \text{span} \times (\text{span}/3) = 25 \times (25/3) = 208 \text{ ft}^2 \ (19.4 \text{ m}^2) \end{cases}$$

Therefore, $A = 208 \text{ ft}^2$ (19.4 m²) for the interior joists.
For the end joists:

$$A = \text{larger of} \begin{cases} 25 \times [(8/2) + 1.5] = 137.5 \text{ ft}^2 \ (12.8 \text{ m}^2) \\ \\ \text{span} \times (\text{span}/3) = 25 \times (25/3) = 208 \text{ ft}^2 \ (19.4 \text{ m}^2) \end{cases}$$

Therefore, $A = 208 \text{ ft}^2$ (19.4 m²) for the end joists.

Zone	(GC_p)	p, lb/ft² (kN/m²)
4	0.79	21.9 (1.05)
	−0.88	−24.0 (−1.15)
5	0.79	21.9 (1.05)
	−1.04	−27.6 (−1.32)

TABLE 6.11   Design Wind Pressures, *p*, on the Walls in Example 6.1

	(GC$_p$)	
Zone	Positive	Negative
1′	0.20	−0.74
1	0.20	−1.16
2	0.20	−1.60
3	0.20	−1.80

TABLE 6.12 External Pressure Coefficients, (GC$_p$), for the Roof in Example 6.1

Step 3—Determine the external pressure coefficients, (GC$_p$)
ASCE/SEI Figure 30.3-2A

The external pressure coefficients on the roof are given in Table 6.12 based on a roof without overhangs. The equations in ASCE/SEI Table C30.3-2 can be used to determine the external pressure coefficients.

Step 4—Determine the internal pressure coefficients, (GC$_{pi}$)
Table 2.9

From Step 7 in Example 3.1, $(GC_{pi}) = +0.18, -0.18$ for an enclosed building.

Step 5—Determine the design wind pressures, p
ASCE/SEI Equation (30.3-1)

$$p = q_h[(GC_p) - (GC_{pi})]$$

The maximum design wind pressures on the roof joists are given in Table 6.13. Design wind pressures for zone 3 are as follows:

- For positive (GC_p): $p = 22.6 \times [0.20 - (-0.18)] = 8.6 \text{ lb/ft}^2$
- For negative (GC_p): $p = 22.6 \times [(-1.80) - (+0.18)] = -44.8 \text{ lb/ft}^2$

Zone	(GC$_p$)	p, lb/ft² (kN/m²)
1′	0.20	8.6 (0.41)
	−0.74	−20.8 (0.99)
1	0.20	8.6 (0.41)
	−1.16	−30.3 (1.45)
2	0.20	8.6 (0.41)
	−1.60	−40.2 (1.92)
3	0.20	8.6 (0.41)
	−1.80	−44.8 (2.14)

TABLE 6.13 Design Wind Pressures, p, on the Roof Joists in Example 6.1

In S.I.:

- For positive (GC_p): $p = 1.08 \times [0.20 - (-0.18)] = 0.41 \text{ kN/m}^2$
- For negative (GC_p): $p = 1.08 \times [(-1.80) - (+0.18)] = -2.14 \text{ kN/m}^2$

The positive design wind pressures in all zones must be increased to the minimum pressure of 16 lb/ft² (0.77 kN/m²) in accordance with ASCE/SEI 30.2.2.

The maximum wind pressures are applied normal to the joists and act over the tributary area of each joist, which is equal to 200 ft² (18.6 m²) for an interior joist and 137.5 ft² (12.8 m²) for an end joist. If the tributary area were greater than 700 ft² (65.0 m²), the joists could be designed using the requirements for MWFRSs (ASCE/SEI 30.2.3).

Loading diagrams for typical roof joists are given in Fig. 6.14. Based on the widths of the designated zones in ASCE/SEI Figure 30.3-2A, it is evident zone 1' is not applicable [this corresponds with ASCE/SEI Figure C30-1: the least horizontal dimension of the building, which is equal to 50 ft (15.2 m), is greater than $1.2h = 43.2$ ft (13.2 m) and is less than $2.4h = 86.4$ ft (26.3 m)].

6.11.2   Example 6.2—Design Wind Pressures on C&C Elements of a Commercial Building, Chapter 30, Part 2

Determine the design wind pressures on (1) the 7-in. (178-mm) precast concrete wall and (2) the roof joists for the commercial building in Example 3.1 (see Fig. 3.13) using Part 2 in Chapter 30 and the design data Table 3.2. The roof joists are spaced 8 ft (2.4 m) on center.

Solution
Check if the building meets all the conditions and limitations in ASCE/SEI 30.1.1, 30.1.2, and 30.4.1 so that Part 2 in Chapter 30 can be used to determine the design wind pressures on the C&C elements:

- The building is enclosed (see Table 3.2).
- The building is a low-rise building because $h \leq 60$ ft (18.3 m) and is less than the minimum horizontal dimension.
- The building has a flat roof.
- The building is regular-shaped, that is, the structure has no unusual geometrical irregularities in spatial form.
- The building does not have response characteristics that make it subject to across-wind loading, vortex shedding, or instability caused by galloping or flutter. Additionally, the building is not located at a site where channeling effects or buffeting in the wake of upwind obstructions warrant special consideration.

Therefore, the conditions in ASCE/SEI 30.1.1, 30.1.2, and 30.4.1 are met and the requirements in Part 2 of Chapter 30 may be used.

The flowchart in Fig. 6.3 is used to determine the design wind pressures, p_{net}, on the C&C elements.

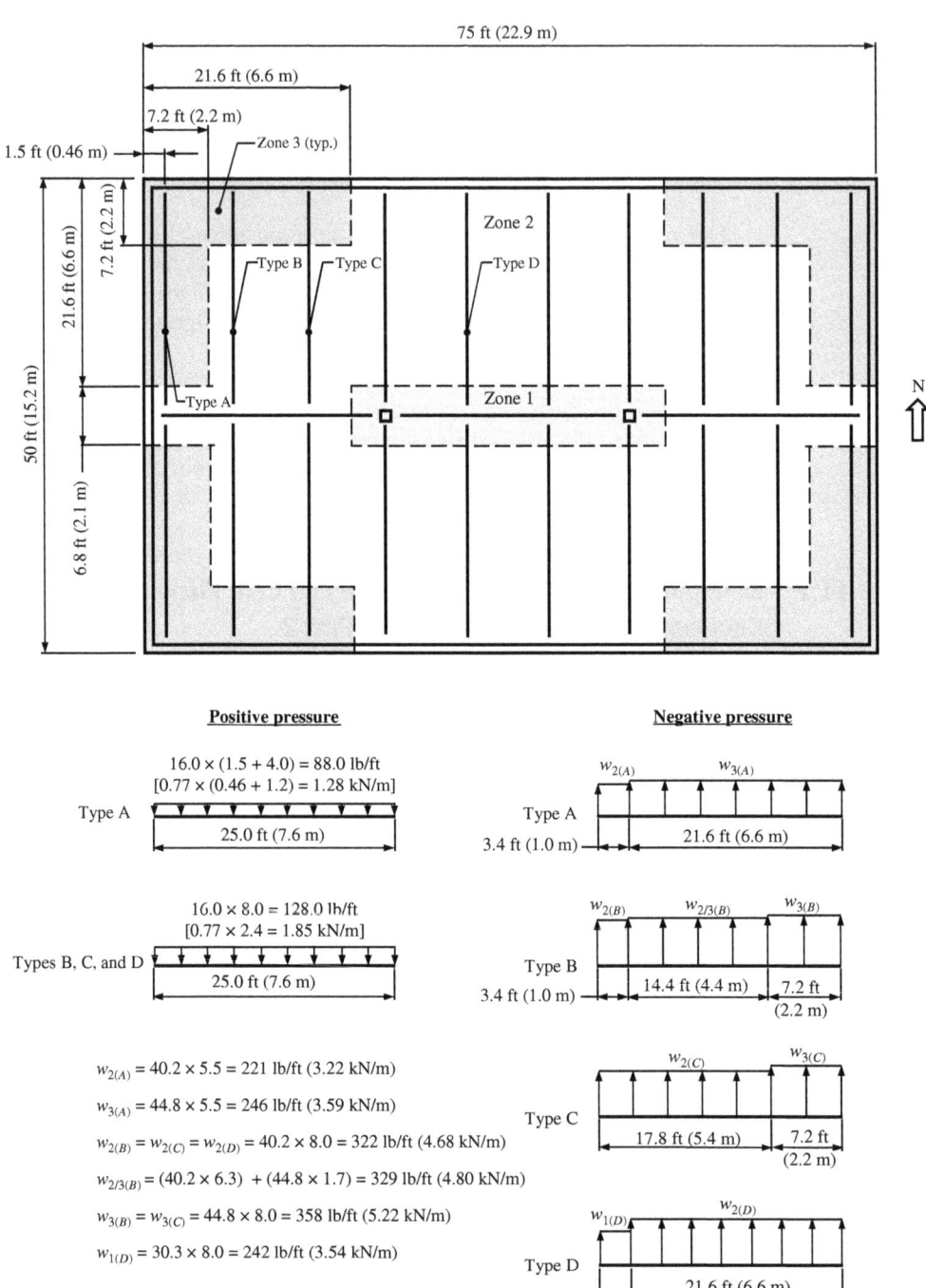

FIGURE 6.14 Joist loading diagrams for Example 6.1.

Part 1—Determine the design wind pressures on the 7-in. (178-mm) precast concrete wall

> *Step 1—Determine the surface roughness category*   ASCE/SEI 26.7.2
>
> The surface roughness category is given as C in the design data (see Table 3.2).
>
> *Step 2—Determine the exposure category*   ASCE/SEI 26.7.3
>
> It is assumed that surface roughness C applies in all directions and that exposures B and D are not applicable. Therefore, the exposure category is C.
>
> *Step 3—Determine the topographic factor, K_{zt}*   ASCE/SEI 26.8
>
> Because the building is not located on a hill, ridge, or escarpment, $K_{zt} = 1.0$.
>
> *Step 4—Determine the risk category of the building*   ASCE/SEI Table 1.5-1
>
> Due to the nature of its occupancy, this commercial building falls under Risk Category II.
>
> *Step 5—Determine the basic wind speed, V*   Table 2.1
>
> For Risk Category II, use IBC Figure 1609.3(1) or ASCE/SEI Figure 26.5-1B. Equivalently, use Ref. 3 or Ref. 4 to obtain $V = 101$ mi/h (45 m/s) for Phoenix, AZ.
>
> *Step 6—Determine the effective wind area, A*   Sec. 6.3
>
> The effective wind area for the precast concrete wall, which is supported at the ground and at the floor and roof diaphragms, is determined as follows:
>
> $$A = \text{span} \times (\text{span}/3) = 12 \times (12/3) = 48 \text{ ft}^2 \ (4.5 \text{ m}^2)$$
>
> *Step 7—Determine p_{net30}*   ASCE/SEI Figure 30.4-1
>
> Wind pressures, p_{net30}, are read directly from ASCE/SEI Figure 30.4-1 for a basic wind speed and an effective wind area based on Exposure B and a mean roof height of 30 ft (9.1 m).
>
> According to note 4 in ASCE/SEI Figure 30.4-1, tabulated pressures may be interpolated for effective wind areas between those given in the figure, or the pressure associated with the lower effective area may be used. The former of these options is used in this example. Wind pressures, p_{net30}, are given in Table 6.14 and are obtained by linear interpolation for A and V.
>
> *Step 8—Determine the adjustment factor, λ, for height and exposure*
>   ASCE/SEI Figure 30.4-1
>
> For Exposure C and a mean roof height of 36 ft (11.0 m), $\lambda = 1.46$ by linear interpolation.

Zone	p_{net30}, lb/ft² (kN/m²)	
4	16.5 (0.79)	−18.1 (−0.87)
5	16.5 (0.79)	−20.9 (−1.00)

TABLE 6.14 Wind Pressures, p_{net30}, on the Walls in Example 6.2

Zone	p_{net}, lb/ft² (kN/m²)	
4	24.1 (1.15)	−26.4 (−1.27)
5	24.1 (1.15)	−30.5 (−1.46)

TABLE 6.15   Design Wind Pressures, p_{net}, on the Walls in Example 6.2

Step 9—Determine the net design wind pressures, p_{net}   ASCE/SEI Equation (30.4-1)

$$p_{net} = \lambda K_{zt} p_{net30} = 1.46 \times 1.0 \times p_{net30} = 1.46 p_{net30}$$

The net wind pressures on the wall are given in Table 6.15.
The width of zone 5, a, is determined as follows:

$$a = \text{smaller of} \begin{cases} 0.1 \times \text{least horizontal dimension} = 0.1 \times 50 = 5 \text{ ft (1.5 m)} \\ 0.4h = 0.4 \times 36 = 14.4 \text{ ft (4.4 m)} \end{cases}$$

$$\text{Minimum } a = \text{greater of} \begin{cases} 0.04 \times \text{least horizontal dimension} = 0.04 \times 50 = 2 \text{ ft (0.61 m)} \\ 3 \text{ ft (0.91 m)} \end{cases}$$

Therefore, $a = 5$ ft (1.5 m).

Part 2—Determine the design wind pressures on the roof joists

Step 1—Determine the surface roughness category   ASCE/SEI 26.7.2

The surface roughness category is given as C in the design data (see Table 3.2).

Step 2—Determine the exposure category   ASCE/SEI 26.7.3

It is assumed that surface roughness C applies in all directions and that exposures B and D are not applicable. Therefore, the exposure category is C.

Step 3—Determine the topographic factor, K_{zt}   ASCE/SEI 26.8

Because the building is not located on a hill, ridge, or escarpment, $K_{zt} = 1.0$.

Step 4—Determine the risk category of the building   ASCE/SEI Table 1.5-1

Due to the nature of its occupancy, this commercial building falls under Risk Category II.

Step 5—Determine the basic wind speed, V   Table 2.1

For Risk Category II, use IBC Figure 1609.3(1) or ASCE/SEI Figure 26.5-1B.
Equivalently, use Ref. 3 or Ref. 4 to obtain $V = 101$ mi/h (45 m/s) for Phoenix, AZ.

Step 6—Determine the effective wind area, A Sec. 6.3

The effective wind area for a typical interior roof joist is determined as follows (see Fig. 3.13):

$$A = \text{larger of} \begin{cases} 25 \times 8 = 200 \text{ ft}^2 \ (18.6 \text{ m}^2) \\ \text{span} \times (\text{span}/3) = 25 \times (25/3) = 208 \text{ ft}^2 \ (19.4 \text{ m}^2) \end{cases}$$

Therefore, $A = 208$ ft² (19.4 m²) for the interior joists.

For the end joists:

$$A = \text{larger of} \begin{cases} 25 \times [(8/2)+1.5] = 137.5 \text{ ft}^2 \ (12.8 \text{ m}^2) \\ \text{span} \times (\text{span}/3) = 25 \times (25/3) = 208 \text{ ft}^2 \ (19.4 \text{ m}^2) \end{cases}$$

Therefore, $A = 208$ ft² (19.4 m²) for the end joists.

Step 7—Determine p_{net30} ASCE/SEI Figure 30.4-1

Wind pressures, p_{net30}, are read directly from ASCE/SEI Figure 30.4-1 for a basic wind speed and an effective wind area based on Exposure B and a mean roof height of 30 ft (9.1 m). Because the effective wind area is greater than 100 ft² (30.5 m²) for the joists, the tabulated wind pressures associated with an effective wind area of 100 ft² (30.5 m²) are applicable.

Wind pressures, p_{net30}, are given in Table 6.16 and are obtained by linear interpolation for *V*. Based on the widths of the designated zones in ASCE/SEI Figure 30.4-1, it is evident zone 1′ is not applicable [this corresponds with ASCE/SEI Figure C30-1: the least horizontal dimension of the building, which is equal to 50 ft (15.2 m), is greater than $1.2h = 43.2$ ft (13.2 m) and is less than $2.4h = 86.4$ ft (26.3 m)].

Step 8—Determine the adjustment factor, λ, for height and exposure
ASCE/SEI Figure 30.4-1

For Exposure C and a mean roof height of 36 ft (11.0 m), $\lambda = 1.46$ by linear interpolation.

Step 9—Determine the net design wind pressures, p_{net} ASCE/SEI Equation (30.4-1)

$$p_{net} = \lambda K_{zt} p_{net30} = 1.46 \times 1.0 \times p_{net30} = 1.46 p_{net30}$$

The net wind pressures on the wall are given in Table 6.17.

Zone	p_{net30}, lb/ft² (kN/m²)	
1	5.9 (0.28)	−22.9 (−1.10)
2	5.9 (0.28)	−30.3 (−1.45)
3	5.9 (0.28)	−36.1 (−1.73)

TABLE 6.16 Wind Pressures, p_{net30}, on the Roof Joists in Example 6.2

Zone	p_{net}, lb/ft² (kN/m²)	
1	8.6 (0.41)	−33.4 (−1.61)
2	8.6 (0.41)	−44.2 (−2.12)
3	8.6 (0.41)	−52.7 (−2.53)

TABLE 6.17 Design Wind Pressures, p_{net}, on the Roof Joists in Example 6.2

The positive design wind pressures in all zones must be increased to the minimum pressure of 16 lb/ft² (0.77 kN/m²) in accordance with ASCE/ SEI 30.2.2.

The maximum wind pressures are applied normal to the joists and act over the tributary area of each joist, which is equal to 200 ft² (18.6 m²) for an interior joist and 137.5 ft² (12.8 m²) for an end joist. If the tributary area were greater than 700 ft² (65.0 m²), the joists could be designed using the requirements for MWFRSs (ASCE/SEI 30.2.3).

Loading diagrams for typical roof joists are given in Fig. 6.15.

6.11.3 Example 6.3—Design Wind Pressures on C&C Elements of a Residential Building, Chapter 30, Part 3

Determine the design wind pressures on the C&C elements of the walls for the residential building in Example 3.6 (see Fig. 3.26) using Part 3 in Chapter 30 and the design data Table 3.21. The glazing system consists of 5-ft (1.5-m) wide by 8-ft (2.4-m) deep glazing panels and 8-ft (2.4-m) long mullions (see Fig. 6.1).

Solution
Check if the building meets all the conditions and limitations in ASCE/SEI 30.1.1, 30.1.2, and 30.5.1 so that Part 3 in Chapter 30 can be used to determine the design wind pressures on the C&C elements:

- The building is enclosed (see Table 3.21).
- The mean roof height of the building $h = 152.0$ ft (46.3 m) > 60 ft (18.3 m).
- The building has a flat roof.
- The building is regular-shaped, that is, the structure has no unusual geometrical irregularities in spatial form.
- The building does not have response characteristics that make it subject to across-wind loading, vortex shedding, or instability caused by galloping or flutter. Additionally, the building is not located at a site where channeling effects or buffeting in the wake of upwind obstructions warrant special consideration.

Therefore, the conditions in ASCE/SEI 30.1.1, 30.1.2, and 30.5.1 are met and the requirements in Part 3 of Chapter 30 may be used.

The design procedure in Fig. 6.4 is used to determine the design wind pressures, p, on the C&C elements.

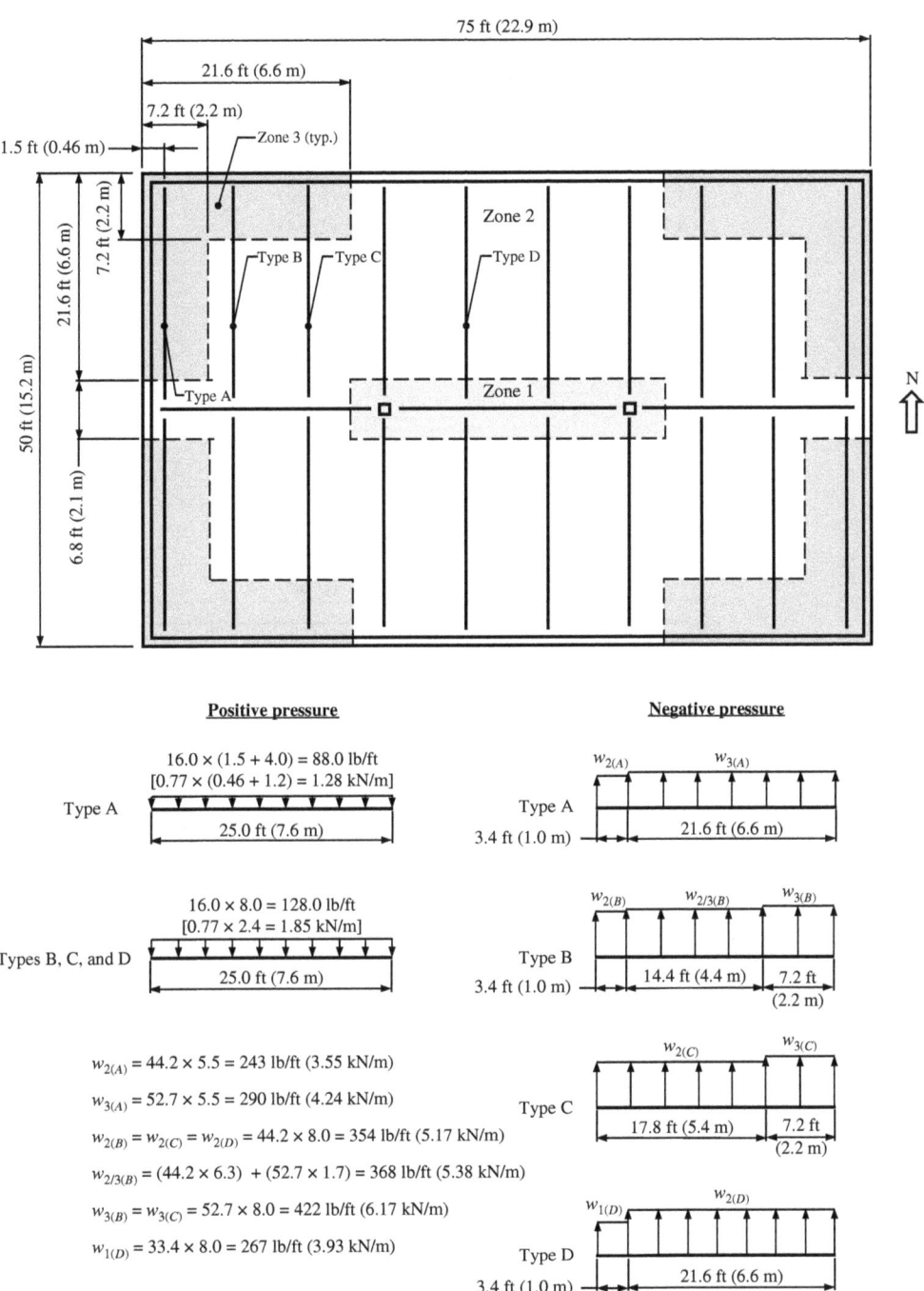

The figure contains the following labels and annotations:

75 ft (22.9 m)

21.6 ft (6.6 m)

7.2 ft (2.2 m)

1.5 ft (0.46 m)

7.2 ft (2.2 m)

21.6 ft (6.6 m)

50 ft (15.2 m)

6.8 ft (2.1 m)

Zone 3 (typ.)

Zone 2

Type B   Type C   Type D

Type A

Zone 1

N

Positive pressure

Type A

$16.0 \times (1.5 + 4.0) = 88.0$ lb/ft
$[0.77 \times (0.46 + 1.2) = 1.28$ kN/m]

25.0 ft (7.6 m)

Types B, C, and D

$16.0 \times 8.0 = 128.0$ lb/ft
$[0.77 \times 2.4 = 1.85$ kN/m]

25.0 ft (7.6 m)

$w_{2(A)} = 44.2 \times 5.5 = 243$ lb/ft (3.55 kN/m)

$w_{3(A)} = 52.7 \times 5.5 = 290$ lb/ft (4.24 kN/m)

$w_{2(B)} = w_{2(C)} = w_{2(D)} = 44.2 \times 8.0 = 354$ lb/ft (5.17 kN/m)

$w_{2/3(B)} = (44.2 \times 6.3) + (52.7 \times 1.7) = 368$ lb/ft (5.38 kN/m)

$w_{3(B)} = w_{3(C)} = 52.7 \times 8.0 = 422$ lb/ft (6.17 kN/m)

$w_{1(D)} = 33.4 \times 8.0 = 267$ lb/ft (3.93 kN/m)

Negative pressure

Type A

$w_{2(A)}$ $w_{3(A)}$

3.4 ft (1.0 m)   21.6 ft (6.6 m)

Type B

$w_{2(B)}$ $w_{2/3(B)}$ $w_{3(B)}$

3.4 ft (1.0 m)   14.4 ft (4.4 m)   7.2 ft (2.2 m)

Type C

$w_{2(C)}$ $w_{3(C)}$

17.8 ft (5.4 m)   7.2 ft (2.2 m)

Type D

$w_{1(D)}$ $w_{2(D)}$

3.4 ft (1.0 m)   21.6 ft (6.6 m)

FIGURE 6.15 Joist loading diagrams for Example 6.2.

Height above Ground Level, z, ft (m)	q_z, lb/ft² (kN/m²)
152.0 (46.3)	27.6 (1.33)
142.5 (43.4)	27.1 (1.31)
133.0 (40.5)	26.6 (1.28)
123.5 (37.6)	26.2 (1.26)
114.0 (34.8)	25.7 (1.24)
104.5 (31.9)	24.9 (1.20)
95.0 (29.0)	24.2 (1.16)
85.5 (26.1)	23.7 (1.14)
76.0 (23.2)	22.7 (1.09)
66.5 (20.3)	21.9 (1.06)
57.0 (17.4)	20.9 (1.01)
47.5 (14.5)	19.9 (0.96)
38.0 (11.6)	18.7 (0.90)
28.5 (8.7)	17.2 (0.83)
19.0 (5.8)	15.4 (0.74)
9.5 (2.9)	14.2 (0.68)

TABLE 6.18   Velocity Pressure, q_z, for the Building in Example 6.3

Step 1—Determine the velocity pressure, q_z, at height z on the windward wall   Fig. 2.3

The velocity pressures, q_z, over the height of the building are determined in Example 3.6 and are given in Table 3.23 (see Table 6.18).

Step 2—Determine the velocity pressure, q_h, at the mean roof height for the leeward and side walls

From Table 6.18, $q_h = 27.6$ lb/ft² (1.33 kN/m²).

Step 3—Determine the effective wind area, A   Sec. 6.3

The effective wind area for the glazing panels and mullions are determined as follows:

- Glazing panels :

$$A = \text{larger of} \begin{cases} 5\times(8/2) = 20 \text{ ft}^2 \text{ (1.9 m}^2) \\ \text{span}\times(\text{span}/3) = 5\times(5/3) = 8.3 \text{ ft}^2 \text{ (0.77 m}^2) \end{cases}$$

$$\text{Mullions} : A = \text{larger of} \begin{cases} 5\times8 = 40 \text{ ft}^2 \text{ (3.7 m}^2) \\ \text{span}\times(\text{span}/3) = 8\times(8/3) = 21.3 \text{ ft}^2 \text{ (2.0 m}^2) \end{cases}$$

Therefore, $A = 20$ ft² (1.9 m²) for the glazing panels and $A = 40$ ft² (3.7 m²) for the mullions.

Zone	(GC_p)			
	Glazing Panels		Mullions	
	Positive	Negative	Positive	Negative
4	0.90	−0.90	0.84	−0.86
5	0.90	−1.80	0.84	−1.63

TABLE 6.19 External Pressure Coefficients, (GC_p), for the Walls in Example 6.3

Step 4—Determine the external pressure coefficients, (GC_p)

ASCE/SEI Figure 30.5-1

The external pressure coefficients for the walls in zones 4 and 5 are given in Table 6.19.

The width of zone 5, a, is determined as follows:

$$a = \text{larger of} \begin{cases} 0.1 \times \text{least horizontal dimension} = 0.1 \times 61 = 6.1 \text{ ft (1.9 m)} \\ 3 \text{ ft (0.91 m)} \end{cases}$$

Therefore, $a = 6.1$ ft (1.9 m).

Step 5—Determine the velocity pressure for internal pressure determination, q_i

According to ASCE/SEI 30.5.2, $q_i = q_h = 27.6$ lb/ft² (1.33 kN/m²)

Step 6—Determine the enclosure classification   Table 2.7

In the design data, the building is given as enclosed.

Step 7—Determine the internal pressure coefficients, (GC_{pi})   Table 2.9

From step 7 in Example 3.6, $(GC_{pi}) = +0.18, -0.18$ for an enclosed building.

Step 8—Determine the design wind pressures, p   ASCE/SEI Equation (30.5-1)

- Windward wall

$$p = q_z(GC_p) - q_i(GC_{pi}) = q_z(GC_p) - 27.6(\pm 0.18) = q_z(GC_p) \mp 5.0 \text{ lb/ft}^2$$

In S.I.:

$$p = q_z(GC_p) - 1.33(\pm 0.18) = q_z(GC_p) \mp 0.24 \text{ kN/m}^2$$

- Leeward and side walls

$$p = q_h(GC_p) - q_i(GC_{pi}) = 27.6(GC_p) - 27.6(\pm 0.18) = 27.6(GC_p) \mp 5.0 \text{ lb/ft}^2$$

In S.I.:

$$p = 1.33(GC_p) - 1.33(\pm 0.18) = 1.33(GC_p) \mp 0.24 \text{ kN/m}^2$$

The maximum design wind pressures for positive and negative internal pressures are given in Table 6.20. The maximum positive pressure, which

Height above Ground Level, z, ft (m)	Design Pressure p, lb/ft² (kN/m²)							
	Glazing Panels				Mullions			
	Zone 4		Zone 5		Zone 4		Zone 5	
	Positive	Negative	Positive	Negative	Positive	Negative	Positive	Negative
152.0 (46.3)	29.8 (1.44)	−29.8 (−1.44)	29.8 (1.44)	−54.7 (−2.63)	28.2 (1.36)	−28.7 (−1.38)	28.2 (1.36)	−50.0 (−2.41)
142.5 (43.4)	29.4 (1.42)	−29.8 (−1.44)	29.4 (1.42)	−54.7 (−2.63)	27.8 (1.34)	−28.7 (−1.38)	27.8 (1.34)	−50.0 (−2.41)
133.0 (40.5)	28.9 (1.39)	−29.8 (−1.44)	28.9 (1.39)	−54.7 (−2.63)	27.3 (1.32)	−28.7 (−1.38)	27.3 (1.32)	−50.0 (−2.41)
123.5 (37.6)	28.6 (1.37)	−29.8 (−1.44)	28.6 (1.37)	−54.7 (−2.63)	27.0 (1.30)	−28.7 (−1.38)	27.0 (1.30)	−50.0 (−2.41)
114.0 (34.8)	28.1 (1.36)	−29.8 (−1.44)	28.1 (1.36)	−54.7 (−2.63)	26.6 (1.28)	−28.7 (−1.38)	26.6 (1.28)	−50.0 (−2.41)
104.5 (31.9)	27.4 (1.32)	−29.8 (−1.44)	27.4 (1.32)	−54.7 (−2.63)	25.9 (1.25)	−28.7 (−1.38)	25.9 (1.25)	−50.0 (−2.41)
95.0 (29.0)	26.8 (1.28)	−29.8 (−1.44)	26.8 (1.28)	−54.7 (−2.63)	25.3 (1.21)	−28.7 (−1.38)	25.3 (1.21)	−50.0 (−2.41)
85.5 (26.1)	26.3 (1.27)	−29.8 (−1.44)	26.3 (1.27)	−54.7 (−2.63)	24.9 (1.20)	−28.7 (−1.38)	24.9 (1.20)	−50.0 (−2.41)
76.0 (23.2)	25.4 (1.22)	−29.8 (−1.44)	25.4 (1.22)	−54.7 (−2.63)	24.1 (1.16)	−28.7 (−1.38)	24.1 (1.16)	−50.0 (−2.41)
66.5 (20.3)	24.7 (1.19)	−29.8 (−1.44)	24.7 (1.19)	−54.7 (−2.63)	23.4 (1.13)	−28.7 (−1.38)	23.4 (1.13)	−50.0 (−2.41)
57.0 (17.4)	23.8 (1.15)	−29.8 (−1.44)	23.8 (1.15)	−54.7 (−2.63)	22.6 (1.09)	−28.7 (−1.38)	22.6 (1.09)	−50.0 (−2.41)
47.5 (14.5)	22.9 (1.10)	−29.8 (−1.44)	22.9 (1.10)	−54.7 (−2.63)	21.7 (1.05)	−28.7 (−1.38)	21.7 (1.05)	−50.0 (−2.41)
38.0 (11.6)	21.8 (1.05)	−29.8 (−1.44)	21.8 (1.05)	−54.7 (−2.63)	20.7 (1.00)	−28.7 (−1.38)	20.7 (1.00)	−50.0 (−2.41)
28.5 (8.7)	20.5 (0.99)	−29.8 (−1.44)	20.5 (0.99)	−54.7 (−2.63)	19.4 (0.94)	−28.7 (−1.38)	19.4 (0.94)	−50.0 (−2.41)
19.0 (5.8)	18.9 (0.91)	−29.8 (−1.44)	18.9 (0.91)	−54.7 (−2.63)	17.9 (0.86)	−28.7 (−1.38)	17.9 (0.86)	−50.0 (−2.41)
9.5 (2.9)	17.8 (0.85)	−29.8 (−1.44)	17.8 (0.85)	−54.7 (−2.63)	16.9 (0.81)	−28.7 (−1.38)	16.9 (0.81)	−50.0 (−2.41)

TABLE 6.20 Design Wind Pressures, p, on the Walls in Example 6.3

varies with height on the windward wall, is obtained with negative internal pressure. The maximum negative pressure, which is constant over the height of the building on the leeward and side walls, is obtained with positive internal pressure. The pressures are applied perpendicular to the C&C elements and act over the respective tributary areas.

The maximum design wind pressures are greater than the minimum pressure of 16 lb/ft² (0.77 kN/m²) acting in either direction normal to the surface (ASCE/SEI 30.2.2).

6.11.4 Example 6.4—Design Wind Pressures on C&C Elements of a Residential Building, Chapter 30, Part 4

Determine the design wind pressures on the C&C elements of the parapet of the residential building in Example 3.6 (see Fig. 3.26) using Part 4 in Chapter 30 and the design data Table 3.21. The height of the parapet is equal to 4.5 ft (1.4 m).

Solution

Check if the building meets all the conditions and limitations in ASCE/SEI 30.1.1, 30.1.2, and 30.6 so that Part 4 in Chapter 30 can be used to determine the design wind pressures on the C&C elements:

- The building is enclosed (see Table 3.21).
- The mean roof height of the building, $h = 152.0$ ft (46.3 m), which is greater than 60 ft (18.3 m) and less than 160 ft (48.8 m).
- The building has a flat roof.
- The building is regular-shaped, that is, the structure has no unusual geometrical irregularities in spatial form.
- The building does not have response characteristics that make it subject to across-wind loading, vortex shedding, or instability caused by galloping or flutter. Additionally, the building is not located at a site where channeling effects or buffeting in the wake of upwind obstructions warrant special consideration.

Therefore, the conditions in ASCE/SEI 30.1.1, 30.1.2, and 30.6 are met and the requirements in Part 4 of Chapter 30 may be used.

The flowchart in Fig. 6.5 is used to determine the design wind pressures, p, on the C&C elements.

Step 1—Determine the surface roughness category   ASCE/SEI 26.7.2

The surface roughness category is given as B in the design data (see Table 3.21).

Step 2—Determine the exposure category   ASCE/SEI 26.7.3

It is assumed that surface roughness B applies in all directions and that exposures C and D are not applicable. Therefore, the exposure category is B.

Step 3—Determine the topographic factor, K_{zt}   ASCE/SEI 26.8

Because the building is not located on a hill, ridge, or escarpment, $K_{zt} = 1.0$.

Step 4—Determine the risk category   ASCE/SEI Table 1.5-1

Due to the nature of its occupancy, this residential building falls under Risk Category II.

Step 5—Determine the basic wind speed, V   Table 2.1

For Risk Category II, use IBC Figure 1609.3(1) or ASCE/SEI Figure 26.5-1B.
    Equivalently, use Ref. 3 or Ref. 4 to obtain $V = 107$ mi/h (48 m/s) for Atlanta, GA.

Step 6—Determine the effective wind area, A   Sec. 6.3

The effective wind area for the parapet is determined as follows:

$$A = \text{span} \times (\text{span}/3) = 4.5 \times (4.5/3) = 6.8 \text{ ft}^2 \ (0.63 \text{ m}^2)$$

Step 7—Determine the wind pressures, p_{table}   ASCE/SEI Table 30.6-2

Wind pressures are calculated at the corners of the parapet where the pressures are the largest; the corresponding wall and roof zones are 5 and 3, respectively.
    Wind pressures, p_{table}, for load cases 1 and 2 are given in Table 6.21 based on a basic wind speed of 110 mi/h (49 m/s), height to the top of the parapet equal to 156.5 ft (47.7 m), and a flat roof; these pressures were obtained by linear interpolation using the values in ASCE/SEI Table 30.6-2.
    Because the basic wind speed in this example is equal to 107 mi/h (48 m/s), the pressures in Table 6.21 can be multiplied by $107^2/110^2 = 0.95$ (see note 2 in ASCE/SEI Table 30.6-2). The adjusted wind pressures are given in Table 6.22.

Step 8—Determine the exposure adjustment factor (EAF)   ASCE/SEI Table 30.6-2

The (EAF) must be determined because the building is assigned to Exposure B.
    From ASCE/SEI Table 30.6-2, (EAF) = 0.808 by linear interpolation for Exposure B and $h = 156.5$ ft (47.7 m).

	p_{table}, lb/ft² (kN/m²)	
Load Case	Zone 3	Zone 5
1	−123.7 (−5.92)	−72.5 (−3.47)
2	17.6 (0.84)	43.2 (2.07)

TABLE 6.21 Wind Pressures, p_{table}, for the Parapet in Example 6.4

	p_{table}, lb/ft² (kN/m²)	
Load Case	Zone 3	Zone 5
1	−117.5 (−5.62)	−68.9 (−3.30)
2	16.7 (0.80)	41.0 (1.97)

TABLE 6.22 Adjusted Wind Pressures, p_{table}, for the Parapet in Example 6.4

Step 9—Determine the effective wind area reduction factor, (RF)

ASCE/SEI Table 30.6-2

Because $A = 6.8$ ft^2 $(0.63$ m$^2) < 10.0$ ft^2 $(0.93$ m$^2)$, $(RF) = 1.0$ for all zones.

Step 10—Determine the design wind pressures, p ASCE/SEI Equation (30.6-1)

$$p = p_{\text{table}}(\text{EAF})(\text{RF})K_{zt} = p_{\text{table}} \times 0.808 \times 1.0 \times 1.0 = 0.808 p_{\text{table}}$$

- Load case A

  The pressures on the windward parapet are equal to the following (see Figure 6.6):

  $$p_1 = p_5 = p_{\text{table}|\text{Zone }5}(\text{EAF})(\text{RF})K_{zt} = 0.808 \times 41.0 = 33.1 \text{ lb/ft}^2$$

  $$p_2 = p_7 = p_{\text{table}|\text{Zone }3}(\text{EAF})(\text{RF})K_{zt} = 0.808 \times (-117.5) = -94.9 \text{ lb/ft}^2$$

  In S.I.:

  $$p_1 = p_5 = 0.808 \times 1.97 = 1.59 \text{ kN/m}^2$$

  $$p_2 = p_7 = 0.808 \times (-5.62) = -4.54 \text{ kN/m}^2$$

- Load case B

  The pressures on the leeward parapet are equal to the following (see Fig. 6.6):

  $$p_3 = p_5 = p_{\text{table}|\text{Zone }5}(\text{EAF})(\text{RF})K_{zt} = 0.808 \times 41.0 = 33.1 \text{ lb/ft}^2$$

  $$p_4 = p_6 = p_{\text{table}|\text{Zone }5}(\text{EAF})(\text{RF})K_{zt} = 0.808 \times (-68.9) = -55.7 \text{ lb/ft}^2$$

  In S.I.:

  $$p_3 = p_5 = 0.808 \times 1.97 = 1.59 \text{ kN/m}^2$$

  $$p_4 = p_6 = 0.808 \times (-3.30) = -2.67 \text{ kN/m}^2$$

These pressures, which are greater than the minimum pressure of 16 lb/ft^2 (0.77 kN/m^2) acting in either direction normal to the surface (ASCE/SEI 30.2.2), are applied to the windward and leeward parapets as shown in Fig. 6.6.

6.11.5 Example 6.5—Design Wind Pressures on C&C Elements of an Open Building, Chapter 30, Part 5

Determine the design wind pressures on the roof joists of the open building in Example 3.7 (see Fig. 3.31) using Part 5 in Chapter 30 and the design data Table 3.28. The joists are spaced 5.0 ft (1.5 m) on center and span 31 ft (9.5 m). Assume obstructed wind flow beneath the roof.

Solution

Check if the building meets all the conditions and limitations in ASCE/SEI 30.1.1, 30.1.2, and 30.7 so that Part 5 in Chapter 30 can be used to determine the design wind pressures on the C&C elements:

- The building is open (see Table 3.28).
- The building has a monosloped free roof.
- The building is regular-shaped, that is, the structure has no unusual geometrical irregularities in spatial form.
- The building does not have response characteristics that make it subject to across-wind loading, vortex shedding, or instability caused by galloping or flutter. Additionally, the building is not located at a site where channeling effects or buffeting in the wake of upwind obstructions warrant special consideration.

Therefore, the conditions in ASCE/SEI 30.1.1, 30.1.2, and 30.7 are met and the requirements in Part 5 of Chapter 30 may be used.

The design procedure in Fig. 6.8 is used to determine the design wind pressures, p, on the C&C elements.

Step 1—Determine the velocity pressure, q_h, at the mean roof height   Fig. 2.3

From step 1 in Example 3.7, $q_h = 23.3$ lb/ft² (1.12 kN/m²).

Step 2—Determine the gust-effect factor   Fig. 2.4

From step 2 in Example 3.7, $G = 0.85$ for rigid buildings.

Step 3—Determine the effective wind area, A   Sec. 6.3

The effective wind area for a typical interior roof joist is determined as follows:

$$A = \text{larger of} \begin{cases} 31 \times 5 = 155 \text{ ft}^2 \text{ (14.4 m}^2) \\ \\ \text{span} \times (\text{span}/3) = 31 \times (31/3) = 320.3 \text{ ft}^2 \text{ (29.8 m}^2) \end{cases}$$

Therefore, $A = 320.3$ ft² (29.8 m²) for the interior joists.

Step 4—Determine the net pressure coefficient, C_N   ASCE/SEI Figure 30.7-1

Zone width, a, is determined as follows:

$$a = \text{smaller of} \begin{cases} 0.1 \times \text{least horizontal dimension} = 0.1 \times 60 = 6 \text{ ft (1.8 m)} \\ \\ 0.4h = 0.4 \times 28 = 11.2 \text{ ft (3.4 m)} \end{cases}$$

$$\text{Minimum } a = \text{greater of} \begin{cases} 0.04 \times \text{least horizontal dimension} = 0.04 \times 60 = 2.4 \text{ ft (0.73 m)} \\ \\ 3 \text{ ft (0.91 m)} \end{cases}$$

	C_N	
Zone	Positive	Negative
1	1.2	−2.1
2	1.2	−2.1
3	1.2	−2.1

TABLE 6.23   Net Pressure Coefficients, C_N, for the Open Building in Example 6.5

Therefore, $a = 6$ ft (1.8 m).

$$4.0a^2 = 144 \text{ ft}^2 \ (13.4 \text{ m}^2) < A = 320.3 \text{ ft}^2 \ (29.8 \text{ m}^2)$$

Net pressure coefficients, C_N, are given in Table 6.23 based on obstructed wind flow, a roof slope $\theta = 15$ degrees, and $A > 4.0a^2$.

Step 5—Determine the net design wind pressure, p   ASCE/SEI Equation (30.7-1)

$$p = q_h G C_N = 23.3 \times 0.85 \times C_N = 19.8 C_N \text{ lb/ft}^2$$

In S.I.:

$$p = 1.12 \times 0.85 \times C_N = 0.95 C_N \text{ kN/m}^2$$

It is evident from Table 6.23 that the roof pressures are the same in all three zones. Therefore,

$$p = 19.8 \times 1.2 = 23.8 \text{ lb/ft}^2$$

$$p = 19.8 \times (-2.1) = -41.6 \text{ lb/ft}^2$$

In S.I.:

$$p = 0.95 \times 1.2 = 1.14 \text{ kN/m}^2$$

$$p = 0.95 \times (-2.1) = -2.00 \text{ kN/m}^2$$

These pressures act perpendicular to the surface and are applied over the tributary area of the joists.

Loading diagrams for a typical interior roof joist are given in Fig. 6.16.

6.11.6   Example 6.6—Design Wind Pressures on C&C Elements of an Attached Canopy, Chapter 30, Part 6

Determine the design wind pressures on the canopy of the office building in Fig. 6.17 using Part 6 in Chapter 30 and the design data Table 6.24. The cantilevered structural members of the canopy are spaced 3 ft (0.91 m) on center.

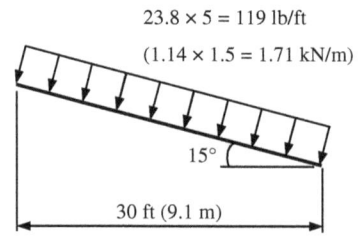

23.8 × 5 = 119 lb/ft

(1.14 × 1.5 = 1.71 kN/m)

15°

30 ft (9.1 m)

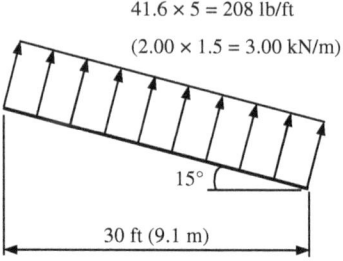

41.6 × 5 = 208 lb/ft

(2.00 × 1.5 = 3.00 kN/m)

15°

30 ft (9.1 m)

FIGURE 6.16   Joist loading diagrams for Example 6.5.

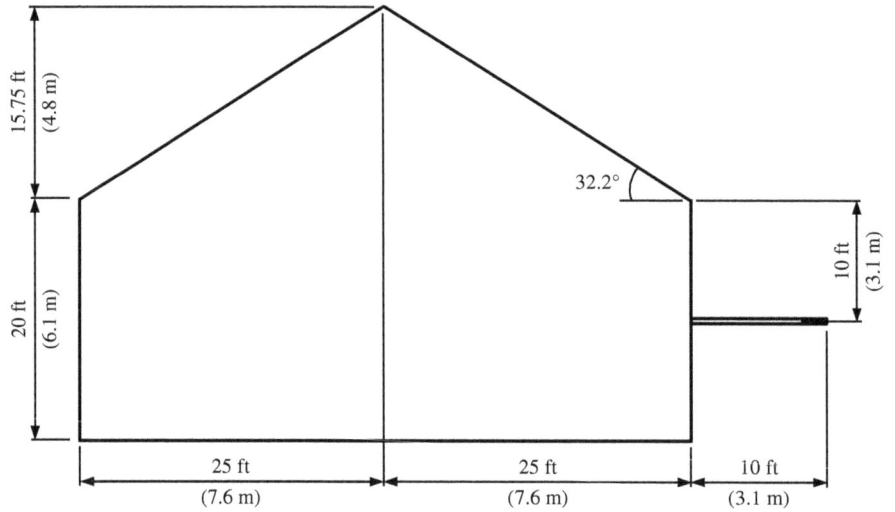

15.75 ft (4.8 m)

20 ft (6.1 m)

32.2°

10 ft (3.1 m)

25 ft (7.6 m)

25 ft (7.6 m)

10 ft (3.1 m)

FIGURE 6.17   Elevation of the office building in Example 6.6.

Location	New Orleans, LA
Surface roughness	C
Topography	Not situated on a hill, ridge, or escarpment
Occupancy	Less than 300 people congregate in one area at the same time
Enclosure classification	Enclosed (site is in a hurricane-prone region as defined in ASCE/SEI 26.2; glazed openings are protected against wind-borne debris in accordance with ASCE/SEI 26.12.3)

TABLE 6.24   Design Data for the Office Building in Example 6.6

Solution

Check if the building meets all the conditions and limitations in ASCE/SEI 30.1.1, 30.1.2, and 30.11 so that Part 6 in Chapter 30 can be used to determine the design wind pressures on the C&C elements:

- The building is regular-shaped, that is, the structure has no unusual geometrical irregularities in spatial form.

- The building does not have response characteristics that make it subject to across-wind loading, vortex shedding, or instability caused by galloping or flutter. Additionally, the building is not located at a site where channeling effects or buffeting in the wake of upwind obstructions warrant special consideration.

- The mean roof height of the building is less than 60 ft (18.3 m).

Therefore, the conditions in ASCE/SEI 30.1.1, 30.1.2, and 30.11 are met and the requirements in Part 6 of Chapter 30 may be used.

The design procedure in Fig. 6.12 is used to determine the design wind pressures, p, on the C&C elements.

Step 1—Determine the velocity pressure, q_h, at the mean roof height   Fig. 2.3

The flowchart in Fig. 2.3 is used to determine q_h.

- Step 1a—Determine the surface roughness category   ASCE/SEI 26.7.2

In Table 6.24, the surface roughness is given as C.

- Step 1b—Determine the exposure category   ASCE/SEI 26.7.3

It is assumed that surface roughness C applies in all directions and that exposures B and D are not applicable. Therefore, the exposure category is C.

- Step 1c—Determine the terrain exposure constants
ASCE/SEI Table 26.11-1

For Exposure C, $\alpha = 9.5$ and $z_g = 900$ ft (274.32 m).

- Step 1d—Determine the velocity pressure exposure coefficient, K_h
ASCE/SEI Table 26.10-1

Mean roof height $(20.0+35.75)/2 = 27.9$ ft (8.5 m)

$$K_h = 2.01(z/z_g)^{2/\alpha} = 2.01\times(27.9/900)^{2/9.5} = 0.97$$

In S.I.:

$$K_h = 2.01\times(8.5/274.32)^{2/9.5} = 0.97$$

- Step 1e—Determine the topographic factor, K_{zt} ASCE/SEI 26.8

Because the building is not located on a hill, ridge, or escarpment, $K_{zt} = 1.0$.

- Step 1f—Determine the wind directionality factor, K_d
ASCE/SEI Table 26.6-1

For the C&C of a building structure, $K_d = 0.85$.

- Step 1g—Determine the ground elevation factor, K_e ASCE/SEI 26.9

  Ground elevation factor can be taken as 1.0 for all elevations.

- Step 1h—Determine the risk category of the building

  ASCE/SEI Table 1.5-1

  Due to the nature of its occupancy, this commercial building falls under Risk Category II.

- Step 1i—Determine the basic wind speed, V Table 2.1

  For Risk Category II, use IBC Figure 1609.3(1) or ASCE/SEI Figure 26.5-1B.

  Equivalently, use Ref. 3 or Ref. 4 to obtain $V = 144$ mi/h (64 m/s) for New Orleans, LA.

- Step 1j—Determine the wind velocity pressure, q_h ASCE/SEI 26.10.2

$$q_h = 0.00256 K_h K_{zt} K_d K_e V^2$$
$$= 0.00256 \times 0.97 \times 1.0 \times 0.85 \times 1.0 \times 144^2 = 43.8 \text{ lb/ft}^2$$

In S.I.:

$$q_h = 0.613 K_h K_{zt} K_d K_e V^2$$
$$= 0.613 \times 0.97 \times 1.0 \times 0.85 \times 1.0 \times 64^2 / 1,000 = 2.07 \text{ kN/m}^2$$

Step 2—Determine the effective wind area, A   Sec. 6.3

The effective wind area for a typical interior cantilevered structural member in the canopy is determined as follows:

$$A = \text{larger of} \begin{cases} 3 \times 10 = 30 \text{ ft}^2 \ (2.8 \text{ m}^2) \\ \text{span} \times (\text{span}/3) = 10 \times (10/3) = 33.3 \text{ ft}^2 \ (3.1 \text{ m}^2) \end{cases}$$

Therefore, $A = 33.3$ ft² (3.1 m²) for a typical cantilevered structural member.

Step 3—Determine the net pressure coefficients, (GC_p)   ASCE/SEI Figure 30.11-1B

The net pressure coefficients, (GC_p), in ASCE/SEI Figure 30.11-1B are used to determine the wind pressures on the cantilevered structural members of the canopy.

$$h_c/h_e = (20.0 - 10.0)/20.0 = 0.5$$

In S.I.:

$$h_c/h_e = (6.1 - 3.05)/6.1 = 0.5$$

For $A = 33.3$ ft² (3.1 m²) and $h_c/h_e = 0.5$, $(GC_p) = 0.80, -0.50$.

Step 4—Determine the design wind pressures, p   ASCE/SEI Equation (30.11-1)

$$p = q_h (GC_p) = \begin{cases} 43.8 \times 0.80 = 35.0 \text{ lb/ft}^2 \\ 43.8 \times (-0.50) = -21.9 \text{ lb/ft}^2 \end{cases}$$

In S.I.:

$$p = q_h(GC_p) = \begin{cases} 2.07 \times 0.80 = 1.66 \text{ kN/m}^2 \\ 2.07 \times (-0.50) = -1.04 \text{ kN/m}^2 \end{cases}$$

These pressures, which are greater than the minimum pressure of 16 lb/ft^2 (0.77 kN/m^2) prescribed in ASCE/SEI 30.2.2, are applied perpendicular to the surface of the canopy (see Fig. 6.18).

6.11.7 Example 6.7—Design Wind Pressures on C&C Elements of an Isolated Circular Tank, Chapter 30, Part 7

Determine the design wind pressures on the isolated circular tank in Example 5.4 (see Fig. 5.13) using Part 7 in Chapter 30 and the design data Table 5.12.

Solution
Check if the structure meets all the conditions and limitations in ASCE/SEI 30.1.1, 30.1.2, and 30.12 so that Part 7 in Chapter 30 can be used to determine the design wind pressures on the C&C elements:

- The structure is regular-shaped, that is, the structure has no unusual geometrical irregularities in spatial form.

- The structure does not have response characteristics that make it subject to across-wind loading, vortex shedding, or instability caused by galloping or flutter. Additionally, the structure is not located at a site where channeling effects or buffeting in the wake of upwind obstructions warrant special consideration.

- The mean roof height of the structure is less than 120 ft (36.6 m).

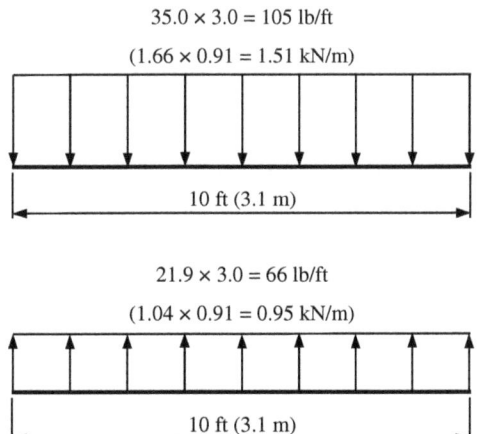

35.0 × 3.0 = 105 lb/ft

(1.66 × 0.91 = 1.51 kN/m)

10 ft (3.1 m)

21.9 × 3.0 = 66 lb/ft

(1.04 × 0.91 = 0.95 kN/m)

10 ft (3.1 m)

FIGURE 6.18 Load diagrams for the structural members of the canopy in Example 6.6.

Therefore, the conditions in ASCE/SEI 30.1.1, 30.1.2, and 30.12 are met and the requirements in Part 7 of Chapter 30 may be used.

The flowchart in Fig. 6.13 is used to determine the design wind pressures, p, on the C&C elements.

Step 1—Determine the velocity pressure, q_h, at the mean roof height   Fig. 2.3

From step 1 in Example 5.4, $q_h = 34.7$ lb/ft^2 (1.68 kN/m^2).

Step 2—Determine the external pressure coefficients, (GC_p), and the internal pressure coefficients, (GC_{pi})

External Pressure Coefficients

- External walls

    External pressure coefficients for external walls of isolated tanks are given in ASCE/SEI Figure 30.12-1.

$$H/D = 15.0/30.0 = 0.5$$ Fig. 5.13

In S.I.:

$$H/D = 4.6/9.1 = 0.5$$

For a wind angle $\alpha = 0$ degrees and an aspect ratio $H/D = 0.5, (GC_p) = 1.00$.

For a wind angle $\alpha = 75$ or 90 degrees and an aspect ratio $H/D = 0.5$, $(GC_p) = -1.10$.

- Roof

    External pressure coefficients for roofs of isolated tanks are given in ASCE/SEI Figure 30.12-2.

    For tanks with flat roofs, zones 1, 2, and 3 are applicable.

    Because the framing for the roof structure is not given, the external pressure coefficients are determined based on an effective wind area ≤ 100 ft^2 (9.3 m^2) (see Table 6.25).

    Width of zone 1: $b = 0.5D = 15.0$ ft (4.6 m) for $H/D = 0.5$ (see ASCE/SEI Figure 30.12-2).

    Width of zone 3: $a = 0.1D = 3.0$ ft (0.91 m).

	(GC_p)	
Zone	**Positive***	**Negative**
1	0.30	−0.80
2	0.30	−0.50
3	0.30	−1.20

*For roof angles less than 10 degrees (line A in ASCE/SEI Figure 30.12-2).

TABLE 6.25   External Pressure Coefficients, (GC_p), for the Roof of the Isolated Circular Tank in Example 6.7

Element		p, lb/ft^2 (kN/m^2)	
		Positive	**Negative**
External walls		41.0 (1.98)*	−44.5 (−2.15)**
Roof	Zone 1	16.7 (0.80)	−34.1 (−1.64)
	Zone 2	16.7 (0.80)	−23.7 (−1.14)
	Zone 3	16.7 (0.80)	−47.9 (−2.32)
Underside	Zones 1 and 2	34.1 (1.64)	−27.1 (−1.31)
	Zone 3	47.9 (2.32)	−37.5 (−1.81)

*$\alpha = 0$ degrees
**$\alpha = 75$ or 90 degrees

TABLE 6.26   Maximum Design Wind Pressures for the Isolated Circular Tank in Example 6.7

- Underside

  External pressure coefficients for the underside of isolated tanks are given in ASCE/SEI 30.12.5.

  For zones 1 and 2 in ASCE/SEI Figure 30.12-2: $(GC_p) = 0.80, -0.60$
  For zone 3 in ASCE/SEI Figure 30.12-2: $(GC_p) = 1.20, -0.90$

Internal Pressure Coefficients

For an enclosed circular tank that is not open-topped, $(GC_{pi}) = 0.18, -0.18$.

Step 3—Determine the design wind pressures, p   ASCE/SEI Equation (30.12-1)

$$p = q_h[(GC_p) - (GC_{pi})] = 34.7[(GC_p) - (\pm 0.18)] = 34.7(GC_p) \mp 6.3 \text{ lb/ft}^2$$

In S.I.:

$$p = 1.68[(GC_p) - (\pm 0.18)] = 1.68(GC_p) \mp 0.30 \text{ kN/m}^2$$

Maximum design wind pressures are given in Table 6.26. These design pressures, which are greater than the minimum pressure of 16 lb/ft^2 (0.77 kN/m^2) prescribed in ASCE/SEI 30.2.2, are applied perpendicular to the surfaces of the tank.

CHAPTER 7

References

1. International Code Council. 2017. *2018 International Building Code*, Washington, DC.
2. Structural Engineering Institute of the American Society of Civil Engineers (ASCE). 2017. *Minimum Design Loads and Associated Criteria for Buildings and Other Structures*, ASCE/SEI 7-16, Reston, VA.
3. American Society of Civil Engineers (ASCE). 2020. ASCE 7 Hazard Tool. https://asce7hazardtool.online/.
4. Applied Technology Council (ATC). 2020. ATC Hazards by Location. https://hazards.atcouncil.org.
5. Structural Engineers Association of California (SEAOC). 2017. *Wind Design for Solar Arrays*, Report SEAOC PV2-2017, Sacramento, CA.

2 04

9 781260 467420